Cyrus Fogg Brackett, And others

Electricity in Daily Life

A Popular Account of the Applications of Electricity to Every Day Uses

Cyrus Fogg Brackett, And others

Electricity in Daily Life
A Popular Account of the Applications of Electricity to Every Day Uses

ISBN/EAN: 9783337067403

Printed in Europe, USA, Canada, Australia, Japan

Cover: Foto ©berggeist007 / pixelio.de

More available books at **www.hansebooks.com**

ELECTRICITY IN DAILY LIFE

A POPULAR ACCOUNT OF THE APPLICATIONS
OF ELECTRICITY TO EVERY DAY USES

BY

CYRUS F. BRACKETT	HERBERT LAWS WEBB
FRANKLIN LEONARD **POPE**	W. S. HUGHES, U.S.N.
JOSEPH WETZLER	JOHN MILLIS, U.S.A.
HENRY MORTON	**A.** E. KENNELLY
CHARLES L. BUCKINGHAM	M. ALLEN STARR, M.D.

WITH **ONE HUNDRED AND** TWENTY-FIVE ILLUSTRATIONS

LONDON
KEGAN PAUL, TRENCH, TRÜBNER & CO.
Limited
MDCCCXCI

CONTENTS.

THE ELECTRIC RAILWAY OF TO-DAY 62

BY JOSEPH WETZLER,

Editor of the "Electrical Engineer," New York.

ELECTRICITY IN LIGHTING 95

BY HENRY MORTON,

President of the Stevens Institute of Technology.

CONTENTS.

PAGE

ELECTRICITY IN THE HOUSEHOLD...................... 239

BY A. E. KENNELLY.

Electrician of Mr. Thomas A. Edison's Laboratory.

GENERAL ADOPTION OF ELECTRIC HOUSEHOLD APPLIANCES—THE BELL—COMPLI-
CATED ANNUNCIATOR SYSTEMS—AUTOMATIC REGULATION OF TEMPERATURE—
ELECTRIC FIRE-ALARM SYSTEM—A DOOR-OPENER—THE REGULATION OF CLOCKS
—ELECTRIC TIME-DETECTOR—THE TELEPHONE ON LARGE ESTATES—DOMESTIC
USES OF THE INCANDESCENT LIGHT—A MEANS OF DECORATION—ILLUMINATED
FOUNTAINS—CONVENIENT ARRANGEMENT OF LIGHTS AND WIRES—THE HAND
GAS-IGNITER—SMALL MOTORS FOR HOUSEHOLD USES—ELECTRIC FANS, PUMPS,
LAWN-MOWERS, SHOE-POLISHERS—THE PHONOGRAPH—ELECTRIC HORTICULTURE
—HEATING BY ELECTRICITY—PLAN OF WIRING A HOUSE.

ELECTRICITY IN RELATION TO THE HUMAN BODY.. 261

BY M. ALLEN STARR, M.D..

Professor of Nervous Diseases, College of Physicians and Surgeons, New York.

EXTRAVAGANT CLAIMS FOR ELECTRICITY AS A CURATIVE AGENT—THREE FORMS OF
ELECTRICAL ENERGY—FRICTIONAL ELECTRICITY—ITS EFFECT ON THE HUMAN
BODY—NO CURATIVE POWER IN AN ELECTRICAL BREEZE—A GENTLE STIMULANT
TO CIRCULATION—ELECTRICITY AS A " MIND CURE"—VOLTAIC ELECTRICITY—
CATALYTIC EFFECTS OF A CURRENT ON THE HUMAN BODY—AN AID TO NUTRI-
TION—NECESSARY PRECAUTIONS—CATAPHORIC ACTION—HOW COCAINE MAY BE
APPLIED—ELECTROTONIC EFFECTS—USE OF AN INTERRUPTED CURRENT IN PA-
RALYSIS—SENSATIONS PRODUCED—IMPORTANT SERVICE IN LOCALIZING BRAIN
FUNCTIONS—DEATH BY ELECTRICITY—A LESS OFFENSIVE MODE OF EXECUTION
THAN HANGING—FARADISM—WORTHLESS MAGNETIC APPLIANCES—ELECTRICAL
INSTRUMENTS—THE PROBE AND CAUTERY.

LIST OF ILLUSTRATIONS.

FULL-PAGE ILLUSTRATIONS.

ILLUSTRATIONS IN THE TEXT.

ELECTRICITY IN THE SERVICE OF MAN.

By C. F. BRACKETT.

ELECTRICAL phenomena have now come to be such important factors in the daily administration of human affairs that the age in which we are living may, with a certain propriety, be called the age of electricity, just as former ones have been called, respectively, the ages of stone, bronze, and iron.

It may be taken for granted that the curiosity or interest of every reader of this book will prompt him to inquire, if he has not already done so, how the mysterious agent which we call electricity is brought under control and directed so as to perform the almost infinitely varied service which is now exacted of it. In fact, almost every industry and art is either so dependent upon, or influenced by, its application that no one, whatever his pursuit, can ignore them and yet hope to attain a foremost place.

It is the purpose of this chapter to set forth, in a general way, some of the common methods in accordance with which the more important

1

electrical phenomena are produced, the laws which these phenomena reveal, and the principles involved in the measurement of electrical quantities. What I shall have to say will be concerning principles which will be fully applied in the course of chapters which are to follow.

The term *electrical* was first employed in 1600, by Dr. Gilbert, to

OTTO GERICKE Patricius et Reipubl: Magdeburgensis Consul, ejusdemq, ad Univers.Pac.Tract.Monasterij et Osnabrugi Legatus.

P.Audry sculp.

designate the attraction which amber ($\eta\lambda\epsilon\kappa\tau\rho\rho\nu$) and other substances of its class exhibit when rubbed and presented to light bodies, such as bits of pith or paper. This term and its corresponding substantive have been everywhere adopted in reference to the phenomena we are about to consider.

If a piece of amber, or resin, and a piece of glass be rubbed to-

* Sir Humphry Davy was one of the earliest and most successful investigators of the effects of the electric current.

gether and then separated, they are no longer indifferent to each other as before, but each attracts the other. In this condition the bodies are both said to be *electrified*, or *charged* with electricity. **Evidence of this** condition is easily secured by suspending one of the charged bodies so that it can move freely, and then presenting the other. An electric charge may be communicated to bodies which have not been rubbed, on merely bringing them in contact with one which is already electrified. For example, a light ball of pith suspended by a silken thread will be charged by such contact, and it can then serve as an electroscope; that is, it can be employed as a means of detecting the electric condition of any body to which it may be presented.* A light straw, balanced so as to turn freely on a fine point, may serve the same purpose.

If the pith-ball electroscope be presented to one of the two rubbed bodies just mentioned, say the glass, it will be attracted to it, and after remaining in contact with it for a short time, it will be repelled. If, now, it be presented to the other body it will be attracted. The two forces being oppositely directed in the two cases, as respects the charged bodies, we have a sufficient justification for saying that there are two kinds or states of electrification, and it is sometimes said that there are two kinds of electricity. The latter statement, however, must be understood to be only a convenient mode of expression which does not imply any knowledge of the nature of electricity itself.

Electricians have adopted the language of mathematics, and they accordingly speak of one state of electrification as *positive* and of the other as *negative*, making the convention that the electrification, or the charge, which glass presents when rubbed with silk shall be regarded as positive.

When metals, and moist bodies which are not metallic, are held in the hand and rubbed, they do not show any signs of electrification. Such bodies, however, may be electrified by rubbing, if the precaution

---

* Gericke was the first to observe repulsion between electrified bodies, and the inventor of the first electrical machine, about 1660.

be taken to support them by means of glass, resin, or, in short, by any body which can be electrified by friction while held in the hand. A metallic sphere, for example, supported by a glass rod, may be strongly charged by whipping it smartly with a piece of dry flannel.

Suppose two metallic spheres so supported, and joined by means of a metallic wire, as below, while somewhat remote from each other. If one of them be struck a few times with dry flannel, both spheres will be found charged in the same sense. If, now, either of the spheres or the wire which joins them be touched with the finger, the entire electrification of the system disappears. The wire in this case is said to conduct electricity from one sphere to the other, or when touched, it, together with the person of the experimenter, conducts the electricity to the earth. All bodies which can act in this way are called *conductors*. Threads of silk, rods of glass, sealing-wax, and the like cannot act in this way, and accordingly they are called non-conductors or *insulators*.

Insulated Metallic Spheres, showing inductive action.

There remains to be described another way of producing the charged condition of insulated conductors. If one of two insulated metallic spheres be charged by means of friction, and then be brought near the other, the latter will show signs of both electrifications at the same time—the remoter portion being charged in the same sense as the originally charged body. This action of the one body on the other is called *induction*. If while this action is manifest the two spheres be widely separated from each other, the sphere which was originally charged will retain its charge, but the other will not. If the spheres be brought near together again, induction will take place as before. If, when this is done, the sphere which is subject to inductive action be touched with the finger, it will appear to be entirely discharged. On removing it from the influence of the inducing sphere, however, it will be found to be charged in the opposite sense. In this way it may be

Insulated Metallic Spheres, electrified by contact with charged conductor and by conduction.

charged as many times as we please, and the successive charges may be employed for any purpose to which we may wish to apply them. We can thus produce an unlimited amount of electricity without impairing the charge of the inducing sphere. This can only be done, however, by the expenditure of work.

When two bodies are in different electrical conditions, so that an attraction exists between them, they are said to be at different *potentials*, or what is the same thing, there is said to be a difference of potential between them.   If the bodies are conductors, and if they be brought in contact, either directly or mediately, by means of a third conductor, a redistribution of electricity will take place, and they will then be at the same potential.

Difference of potential may be determined by weighing the attraction which a charged metallic plate of known dimensions can exert upon another plate at a definite distance from it.   An arrangement

Sir William Thomson's.   After the plan of Harris.
Electrometers for Measuring Difference of Potential.

suitable to exhibit the method is shown at the right in the illustration. At the left is shown an electrometer, designed to accomplish the same end more conveniently and accurately.   The attracting plates are in the interior of the apparatus, and the force is measured indirectly by means of suitable springs.

Electrical machines are only more or less convenient contrivances for producing great differences of electrical potential by means of friction or inductive action as above.   [See Toepler-Holtz Machine.]

The laws of electrical attraction and repulsion thus far considered may be briefly stated thus: Unlike electricities mutually attract, and like electricities mutually repel. The attractions or repulsions are proportional, directly, to the product of the numbers which denote the quantities of the electricities concerned, and inversely to the square of the number of units which measure the distance between them.

At the very beginning of the present century, Volta, stimulated by

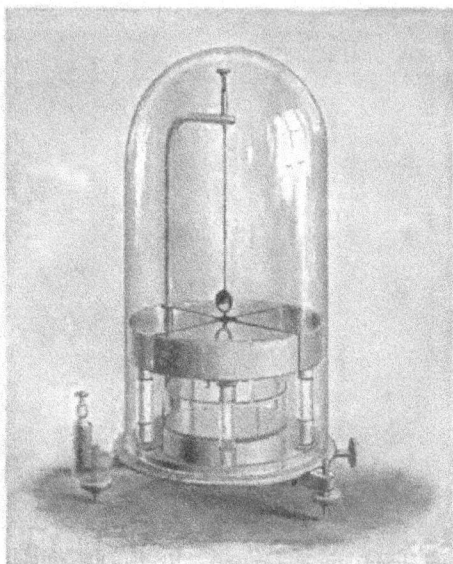

Thomson's Quadrant Electrometer, for comparing potentials.

Galvani's recent discovery of what he called "animal electricity," invented the "pile" and the "crown of cups." We now speak of any equivalent arrangement as a voltaic battery. Without attempting to trace out the path of discovery and invention pursued by Volta, it will be sufficient for our purpose if we make clear the general construction and action of such an apparatus.

If a plate of zinc and a similar one of copper be nearly immersed in water containing a little sulphuric acid, which may be held in any suitable vessel, no noteworthy action will be apparent so long as the metals do not touch; but if they be brought in contact, or be joined by means of a conductor, bubbles of hydrogen gas will at once appear on the surface of the copper, and the zinc will more or less rapidly dissolve to form zinc sulphate with the acid.

If the plates be separated, and the portion of the zinc which remains above the liquid be tested with a very delicate electroscope, it will be

Toepler-Holtz Electrical Machine.

found to be charged with negative electricity, and in like manner the corresponding portion of the copper plate will be found to be charged with positive electricity. These charges are very feeble when compared with those which we can produce by even slightly rubbing a glass rod with a piece of silk. Volta, however, showed that in order to make these charges more evident we have only to combine the actions of several such arrangements as we have just described, by joining the zinc in the first vessel with the copper in the second, and the zinc in the

second with the copper in the third, and so on. In this way the
charges on the terminal plates, or *electrodes*, as they are called, may
be increased to any extent. The difference of potential between the
electrodes is ascribed to the action of a so-called *electromotive force*,

Volta, Inventor of the Voltaic Pile.

arising from the interactions of the different substances employed in the
construction of the battery, and having its analogy in the pressure
which causes liquids to flow along through pipes. If a conducting
wire join the electrodes or terminal plates of metal, a *current of electricity*
will flow through it and through the battery, that is, through the
metals and the liquids, which, with the wire, constitute a closed circuit.
The *intensity* or strength of the current will depend on the magnitude

of the acting electromotive force, and on the *resistance* offered to it by the entire circuit; and investigation shows that it is directly proportional to the former and inversely proportional to the latter. This relation is known as Ohm's law. It is of fundamental importance in both science and engineering.

When a voltaic battery, such as we have described, is put in action by closing its circuit, the intensity of the current rapidly falls off. This is due to the fact that a counter electromotive force is set up, by the hydrogen liberated on the copper plates, which reduces the electromotive force at first acting. This defect, which is common to other forms of voltaic batteries, may be more or less perfectly obviated in various ways. Motion may be given to the plates, whereby the gas will be detached; the plates may be made rough, so as to prevent the strong adhesion of the gas; but it is better to employ some exciting liquid which will not liberate any gaseous product in its action, such as copper sulphate, in the composition of which copper takes the place of hydrogen in the sulphuric acid. In case this salt of copper is employed, it is easy to so arrange the battery that the copper plate shall constantly receive a deposit of bright metallic copper, and so be kept free from adverse action. In the well-known Grove's form of voltaic battery the copper is replaced by platinum, a metal on which the strongest acids do not act, which is placed in a cup of unglazed porcelain containing strong nitric acid. This cup, with its contents, stands in the vessel which contains the zinc plate and the dilute sulphuric acid. The nitric acid is employed to supply oxygen which can unite with the hydrogen as rapidly as it is set free, and thus the platinum plate is kept in the most favorable condition, and counter electromotive force is avoided.

If we wish to study some of the effects which may be produced by means of the battery current, we may employ with advantage a battery of ten or twelve Grove's cups joined in series, that is, the zinc of the first cup joined to the platinum of the second, and so on. A conducting wire should be joined to the platinum of the first cup, and another to the zinc of the last cup.

If the two wires be brought in contact and then separated, a small bright spark will be seen when the contact is broken. The brilliancy of the spark will be much increased if the wires are wrapped around small pencils of hard carbon, and then the latter brought in contact and afterward separated. If the difference of potential between the wires be increased by the employment of a series of cups amounting to forty or fifty, the current will continue to flow even when the pencils of carbon are separated to a distance of two or three millimetres. The carbon pencils will then be heated to an intense whiteness, and a light of dazzling brilliancy will be produced. This is the well-known *arc light* so generally employed in public lighting. The expense and inconvenience attending the use of any form of battery, however, is so great that other means are generally resorted to to supply the electric current, as will presently appear.

If the carbon pencils used for the production of light be replaced by strips of platinum, and if the latter be plunged into water containing about one-tenth its bulk of sulphuric acid, hydrogen will be abundantly liberated from the platinum connected with the zinc electrode of the battery, and oxygen, amounting very exactly to half the volume of the hydrogen, will be liberated from the platinum connected with the platinum electrode. Chemical solutions of the metals may, in like manner, be decomposed by the action of the current. If any conducting body replace the platinum strip connected with the zinc electrode, it may be covered with silver, gold, nickel, or other metal, by employing the proper solution of the metal instead of the acidulated water. This action of the current is called *electrolysis*, and it is largely employed in the arts in the operations of electroplating, electro-metallurgy, etc., as well as in the laboratory in chemical analyses.

Faraday, in the course of a masterly investigation, proved that a given amount of electricity passing through the *electrolyte*, as the solution to be decomposed is called, always sets free a definite amount of its constituents. He also showed that when the current passes through several electrolytes arranged in series, the constituents liberated in any one of them will be proportional to the combining weights of the constituents, respectively.

Various forms of electrolytic apparatus are employed. Those represented in the accompanying illustration are examples.

It will be seen that this law of Faraday gives us the means of comparing one current with another, or of comparing any current with a standard current defined in any way which may be chosen.

If, while the electrical current is flowing through a wire, a delicately poised magnetic needle be carried about it, the needle will tend to place itself at right angles to the general direction of the wire. It may be easily shown that the region about the wire is a magnetic region, commonly called a *magnetic field.* In order to do this, the wire may be made to pass vertically through a sheet of smooth paper which is held in a horizontal position. If, then, while the current is passing, some iron filings be

Two Forms of Electrolytic Apparatus.

a, Battery ; b, silver electrolyte in vessel ; c, apparatus for electrolysis of water.

sprinkled over the paper, and the latter be gently tapped, so as to assist the movement of the filings, they will arrange themselves in concentric circles about the wire. When the current is interrupted the region about the wire is no longer a magnetic field, but it may be restored as often as the current is renewed. This simple experimental fact lies at the foundation of many electrical appliances with which we are familiar. The magnetic field about a single conducting wire is,

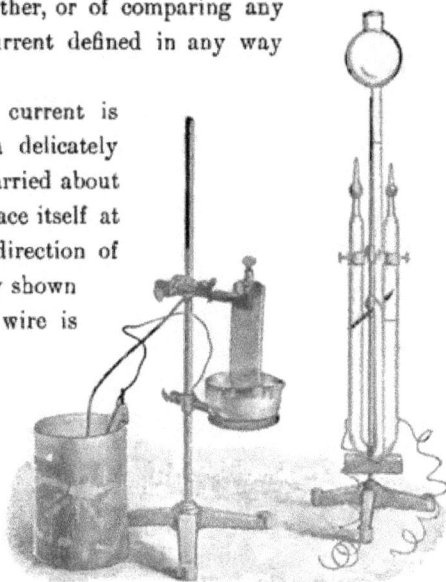

however, generally too feeble to serve for many purposes for which it would otherwise be useful.

In order to strengthen it we may increase the intensity of the current which flows through the conducting wire, or we may employ a sufficient number of conducting wires whose united actions will produce the desired result. The former plan is of limited application, since very intense currents involve great loss of energy in consequence of the heat which they develop, and if too intense they will destroy the conductor. In carrying out the latter plan it is easily seen that it is a matter of indifference whether one employ many separate conductors, each uniting the electrodes of an independent battery, or so dispose a single long conductor that it shall pass many times through the region which it is desired to convert into a magnetic field. The most economical and effective way of proceeding, therefore, is to coil the conductor into a compact helix or spiral. In order that the current shall traverse the entire length of the wire composing the helix, it is covered with an insulating material such as cotton or silk thread. If an open space be left in the centre of the helix, this space, as might be expected, is found, while a current is flowing, to be a powerful magnetic field. If, now, an iron rod be placed in the helix, it is at once powerfully magnetized. Such an apparatus is the well-known electro-magnet which, under one form or another, plays an essential part in a great variety of devices, including the telegraph, the telephone, the burglar-alarm, the dynamo-machine, etc.

Having now a clear idea of some of the more remarkable effects which may be produced by means of the electrical current, we may with advantage consider some additional means of producing the current itself. We have just seen that a bar of iron, when placed within a helix through the wire of which a current of electricity is flowing, becomes a powerful magnet. Experiment shows, conversely, that if the extremities of the wire constituting the helix be disjoined from the battery and brought in contact with each other, and if then a powerful magnet be thrust back and forth in the helix, a current of electricity will be set up which will continue to flow so long only as the motion of the magnet continues. If a bar magnet be carried, in the direction of its length, quite through the helix, the current which is induced in the helix by this motion will be reversed in direction when the mid-point of the magnet passes the mid-point of the length of the helix; and if the magnet be thrust only half way through the helix and then withdrawn, the same result will be produced. It is plain, then, that we can set up an electromotive force in a conductor by merely moving a magnet with reference to the conductor, and that we can determine the direction in which the electromotive force shall act to produce its cor-responding current of electricity by our choice of direction in which the movement shall be made. Moreover, it is quite a matter of in-difference whether we move the magnet or the conductor, or both, so long as the two change their relation to one an-other, and experiment

The Magnetic Field, as indicated by iron filings thrown upon a pane of glass resting upon the poles of an electro-magnet.

shows that the magnitude of the electromotive force, and consequently that of the electrical current, is also under our control. It depends upon the strength of the magnet, on the velocity and direction of its motion, and on the number of turns of wire in the helix.

We might easily devise a machine operated independently, such as the steam-engine, which would continually thrust a magnet into a helix and withdraw it, by a reciprocating motion. We should then have a means of producing currents of electricity which would depend upon mechanical power for the energy which must be supplied. For certain purposes the alternating character of the currents so produced would be a matter of no importance. It is not difficult, however, to devise means by which such a machine can automatically change the relations of **the** conductors, in which the currents flow outside of the helix, so that they shall flow always in the same **direction.** Any form **of such** apparatus is properly called a *magneto-electric machine.*

In practice, machines **of this general character are** usually so constructed as to substitute rotatory **for** reciprocating motion, for the obvious reason that such **a motion is more** easily maintained. In order to carry out **this plan it is only** necessary to place permanently within the **helix a mass of** soft iron, which can readily acquire and lose the **magnetic state on the approach** and recession of a magnet, and then make **a magnet rotate so that its poles shall** pass rapidly in succession near the **mass of iron, or, since it is equally efficient,** make the helix, together with **its iron core, rotate so that it** shall be carried rapidly through **the field** due **to the** magnet. In either case the iron core, **rapidly** acquiring and **losing** its magnetism, **will** act precisely as if it **were a** permanent **magnet having a reciprocating** motion within the helix.

Obviously the permanent **magnet in** the apparatus just considered **may** be replaced by an electro-magnet, and better results may be secured, since such magnets **can** easily be made **of far** greater strength than the **best permanent magnets possess. The** electro-magnet so employed may, **of course, be excited by** the current **supplied** by a voltaic battery; but this inconvenience may be avoided by the simple **device** of sending the current produced by the motion of the **helix** through the coils of the electro-magnet, on whose presence the current itself depends. When this arrangement is adopted the current which is at first produced on setting the apparatus in operation **is** extremely small, since the electromagnet is not then excited to action, and the feeble magnetic field which it presents is wholly due **to** accidental causes. When, however, the least current is produced **by** the motion of the helix, it is made to pass through the coils of the electro-magnet, which has its magnetism developed thereby, and thus presents a more intense field, and this, in **turn, reacts to increase the current.**

Current Produced by Conductor Revolving in the Field of a powerful Electro-magnet.—Effect of Current shown by galvanometer index displaced to right of Scale.

Such are the general principles involved in the construction and operation of the *dynamo-machine.* We need only add that in practice there are several movable coils symmetrically disposed about an axis, and constituting the *armature.* It will thus happen that one or more of the coils will always be passing through the most intense portions of the magnetic field. The current which is generated when the armature revolves is led away to be utilized by means of conducting wires joined to *brushes* or contact-devices, which are suitably held in contact with opposite sides of the revolving armature.

In the dynamo-machine we have an economical means of producing the electric current, since the mechanical energy which must be supplied to it costs less than an equivalent amount of available energy in any other form, and since there is no material contact of the working parts of the machine to wear them out.

The brushes which are applied to the armature are maintained at different potentials when the machine is in action. They may, therefore, be compared to the terminals of the voltaic battery; and, in short, it may be remarked that the relations expressed in Ohm's law hold good for the dynamo-machine and its circuit as well as for the battery.

If the current generated by a dynamo-machine be made to traverse the circuit of another and similar one, the latter will be set in rotation, and it will thus be a means of translating electrical energy into mechanical energy; in other words, it will become an *electric motor,* and may be employed as any other prime motor would be. Since the conductors which unite the machines may be as long as we please, we have thus a means of transferring power from one point to another. Obviously, however, we cannot do this in defiance of the law of conservation of energy, and as there will always be some energy expended in heating the conductors, we can never develop at the distant station as much energy as we expend at the transmitting one.

It is frequently desirable to exchange one current for another of different electromotive force and intensity. The inductive action of the current, through the magnetic field which accompanies its conductor, on other conductors in its neighborhood, affords the means of accomplishing this end. The apparatus by means of which the result is secured is called the *inductorium* or *transformer*. The special arrangement which is employed in any given case depends on the object to be accomplished. The most common form for use in the physical laboratory is shown in the illustration. It consists of two helices of insulated wire of such dimensions that one can be placed within the other. The interior helix has comparatively few turns of thick wire and encloses a bundle of soft iron rods. The exterior one often has many thousand turns of thin wire.

When any change in the strength of the current which is made to pass through the interior helix occurs, a corresponding change takes place in the intensity of the magnetic field dependent upon it, and this, of course, produces a current in the external helix. The electromotive force of this current depends on the number of turns of wire in the helix. We can therefore secure as high electromotive force as we please by adding to the number of turns in the external helix. But every additional turn adds to the resistance and so diminishes the strength of the induced current. If we have given a current of high potential we may send it through the exterior coil, and with every change in its strength we can secure a corresponding current of greater strength but of lower electromotive force.

In practice the current which is to be transformed is rapidly reversed or broken and renewed.

Since, as we have seen, an electric current may be employed to effect chemical decompositions, and since the constituents of an electrolyte so treated may set up a counter-electromotive force, it is clear that we may employ the current produced by the dynamo-machine to set free the constituents of chemical compounds which, on the withdrawal of the decomposing current, will reunite, and in so doing produce a current in the reverse sense. Such an apparatus is called a *storage-battery*. Thus, if two plates of lead which are covered with lead oxide be suspended, without touching, in a vessel containing dilute sulphuric acid, and if,

then, a current from the dynamo-machine be passed through the plates
and through the liquid, the plate connected with the negative electrode
of the machine will give up its oxygen to the liquid, while the plate
connected with the positive electrode will receive from the liquid an
equal amount, in addition to that which it possessed at first. On dis-
connecting the electrolytic apparatus it may itself be employed as a voltaic battery. When conducting-wires from the lead plates are joined,
a current is set up which continues to flow until the plates recover the
condition which they had before the action of the dynamo-current.
Evidently this electrolytic process of *charging* the storage-battery may

Induction Coils, for producing currents of high potential by Induction.

be repeated as frequently as occasion may demand, the battery being in
the mean time used for any purpose to which it is suited.

It remains now to consider, in the most general way, the scheme
which electricians have adopted for the measurement of electrical quantities.

Since we know nothing of the ultimate nature of electricity, but
must admit that it is as much a matter of conjecture as that of gravitation, it is clear that we can only base our measurements upon the effects
which it can produce, just as we are obliged to do when dealing with
the latter agency. Accordingly, we are quite at liberty to express our
results in terms of the same *fundamental units* as are employed in physi-

cal measurements in general, and if we do so we shall obviously be able more easily to detect any relations which **may exist** between electrical and other physical phenomena.

It **may be of** service **to the reader to** define **the** fundamental units **usually** employed, and **to point out how these** enter into the derived **units which must** be employed in measuring complex quantities, such as velocity, acceleration, **force, energy, etc.** When once the method of procedure **in measuring these** quantities is clearly apprehended, there **will be no** difficulty in seeing how we are to apply **it when we have to** measure other related quantities.

Only three fundamental units are needed, and **those** usually selected **are the unit of** *mass* **or** quantity of matter, the unit of *time*, and the unit **of** *length*. **The unit of mass** is the *gram ;* **it is the** one one-thousandth **part of a standard piece of platinum** called the kilogram. Masses are **compared with** copies **of this** unit of mass by means of the balance. The unit of length **is the** *centimetre* [0.394 **inch], or the** one one-hundredth part of **the standard metre.** The unit of **time** is the *second,* or **the one** eighty-six-thousand-four-hundredth part of a mean solar day. **As the** simplest example **of a** derived **unit may** be mentioned that of a surface, **or square centimetre.** It evidently implies extension in **two directions, and into our conception of these extensions the** notion of a **length enters.** We might, however, adopt an arbitrary standard of **surface as our unit, and when we have occasion to measure** a given surface we might find **by actual trial how many times this** standard **unit can be applied to the surface so as to** completely cover every part **of it and no more.** We should thus measure the given surface, but such a measurement would not be **primarily based on a unit of** length, and thus we **should lose sight of important** relations. **In like** manner we might **adopt** arbitrary units of velocity, acceleration, etc., but great difficulties **would** be involved as well as obvious disadvantages.

The notion of velocity implies both a time and a length, and a body moving uniformly in a straight line is said to possess *unit velocity* when **it moves over one unit length—one centimetre—in one second.** If a particle move in a straight line and change its velocity by one unit of velocity in one second, it has one unit *acceleration.* If the particle have a unit of mass, and be subject to unit acceleration, it is acted on by one *unit of force.* Change in the motion of a given mass is thus made the measure of force, and it will be observed that this does not raise any question as to the nature of force itself. The unit of force, as just

Gauss and Weber, who proposed and employed the system of absolute measurements, early in the present century.

defined, is called the *dyne.* It is obvious that a force may be applied to a body which is not free to move by reason of other forces which are acting upon it at the same time. In such cases we have recourse to some indirect method of determining what amount of motion would occur if the body were free to move.

If a body, by reason of its relations to other bodies, is able to exert a force of one dyne, through one centimetre of space, it has one *unit of* energy, called the erg. The energy of a body, or system of bodies, is defined as its ability to perform work. Since, then, energy and work are convertible quantities, it is clear that the erg is the proper unit of measure for work as well as for energy.

A familiar illustration of work is presented when a heavy body is raised from one level to another, as in building; and every reader is aware that in estimating such work both the amount of material raised and the height through which it is raised are taken into account. The work is numerically equal to the product of these two factors. This method of estimating work is sufficiently exact to meet the requirements of the contractor, but not sufficiently so for the purposes of science. The intensity of the earth's attraction is not the same at all points on its surface, and so it follows that the amount of work which must be done in order to raise equal masses through the same height is not everywhere the same. It is not difficult, however, to determine, by means of the pendulum, what acceleration a body falling freely will be subject to at any place. When this is known we can completely specify the force which acts on any given mass by taking the product of the number of units in the mass by the number of units of acceleration. For New York the acceleration due to gravity is about 980 centimetres per second—that is, a body starting from rest and falling freely will acquire a velocity of 980 centimetres in one second. If, therefore, we multiply the number of grams contained in a given body by 980 we have the value of the force acting on the body. This is called, in scientific language, the weight of the body.

Any region in which a mass is acted upon so as to produce, or tend to produce, an acceleration is called a field of gravitational force. In like manner, a region in which a magnet pole is acted upon so as to produce a similar result is called a field of magnetic force, and a region in which an electrified body is so acted upon is called a field of electric force. We must not infer, however, from these forms of expression, or from what we may imagine respecting the regions to which they are applied, that there are several kinds of force. The expressions are in no sense qualitative, but merely indicate the conditions under which a stress will arise such as will tend to produce an acceleration in a body.

Now, the earth itself may be regarded as a great magnet, since it acts on magnets just as they act on each other. The earth is accordingly accompanied by a magnetic field, of which the intensity varies at different points, though not according to any established law. The intensity, however, of the earth's magnetic field at any place can be ascertained by a method in general similar to that employed in determining the intensity of gravitation.

The ratio of the intensities of the two magnetic fields, the one due to the earth, and the other to the action of the current, is determined by the position which a needle, hung free from restraint, assumes under their joint action. But the intensity of the field due to the current depends on its strength, on the distance of the needle from the conductor, and on the number of times the conductor passes through the region. If, then, the conductor be coiled so as to form the circumference of a circle, and have its ends accessible so as to be connected with the source of the current, and if a short magnetic needle be so suspended that its centre of form shall coincide with the centre of the coil, we have all that is essential to a *galvanometer*. Of course, such an instrument would be furnished with a divided circle, or similar device, for convenience in determining the position of the needle when under the influence of the current to be measured. It may easily be shown that the intensities of any two currents are to each other as the tangents of the angles of deflection from the magnetic meridian which they respectively cause the needle to make. Hence the instrument is called a *tangent galvanometer.*

By the aid of the tangent galvanometer and the foregoing principles we can do more than to compare one current with another in terms of the tangents of the angles of deflection which they can cause—we can determine the value of any given current in terms of our fundamental units, or in *absolute measure*, as it is called. In order to this we define the unit magnet pole and the unit current in accordance with

the conventions already made when dealing with velocity, acceleration, etc.

A unit magnet pole is one which exerts a unit force upon a similar
and equal pole at unit distance.

The unit current of electricity, for every centimetre of its length,
can exert a unit force upon a unit magnet pole at one unit distance
from every portion of the current. The only disposition of the conductor conveying the current which will meet the last condition is that
of a circular coil with the needle at its centre. This disposition is
found in the galvanometer described above. As the radius of the coil
described in the definition of unit current would be too small for practice, we have only to employ coils of any convenient, but known, radius,
and make the necessary computations, remembering that the effect of
the current is inversely as the square of its distance from the magnet
pole.

It will be noticed that our definition of current involves but a single
magnet pole, whereas the galvanometer of necessity has a magnetic
needle of two poles, since one pole cannot exist without another having
exactly opposite properties; and, further, nothing is assumed to be
known, in the description of the galvanometer, about the strength of
the poles of the needle employed. These circumstances, however, cause
no inconvenience, since the only use we make of the needle is to ascertain the ratio of the intensity of the horizontal component * of the
earth's magnetic field to the intensity of the field due to the galvanometer coils through which the current to be measured is passing. But the
intensity of the horizontal component of the earth's field is known as
above, in absolute measure, or, in other words, we know the velocity
which it would impart to a free magnet pole weighing one gram. The
intensity of the field due to the current, and therefore the intensity
of the current itself, is then easily found. Thus it is seen that the
intensity of the current is measured in the same fundamental units as are

---

* The direction of the earth's magnetic force is oblique to the horizon. By "horizontal component" is meant the amount of force which it can exert horizontally.

employed in measuring the stress between a heavy body under the
action of gravitation. The real standard of comparison is the horizontal
component of the earth's magnetism. The number representing this, in
the neighborhood of Princeton, is about 0.18 centimetre—in other words,
the velocity along the magnetic meridian which a free magnet pole
weighing one gram would acquire in one second is about 0.18 centimetre. The corresponding velocity which any body would acquire
under the action of gravitation is about nine hundred and eighty centimetres, or about five thousand four hundred and forty-four times greater.

The unit of current, as defined, would necessitate the employment
of inconveniently large numbers in calculations involving the related
quantities, electromotive force, and resistance. To avoid this another
unit, called the *ampère*, is used in practice. It is one-tenth the *electro-
magnetic* unit, as the unit of definition is called.

In order to present an idea of the unit of electromotive force, we
may imagine an experiment of very simple character, although its execution would be very difficult. Let a long straight conductor be bent
upon itself at its middle point, so that the two straight portions shall be at
all other points at unit distance from each other, and let the conductor
thus formed be so fixed in a magnetic field of unit intensity that the
plane in which the conductor lies shall be at right angles to the direction in which a magnetic needle points; if, now, a second conductor
slide along in contact with both branches of the fixed conductor, so
that all positions successively assumed shall be parallel, and with unit
velocity, there will be unit electromotive force set up in the circuit. An
equivalent experiment has been carried out in which a coil of wire of
known dimensions was made to revolve with a known velocity in the
earth's magnetic field. From the data thus at hand the electromotive
force was calculated. It is evidently in terms of the known intensity of
the earth's magnetism at the place of the experiment and of velocity.

Since there is an electromotive force set up in the revolving coil,
there will be a current in it which may be measured by making suitable connections of its terminals with the galvanometer. Both the cur-

rent and the electromotive force being thus known, the resistance of the entire circuit is readily found by Ohm's law. It is simply the ratio of the electromotive force to the current.

The electro-magnetic units of electromotive force and resistance defined as above are too small for practical use. The *practical* unit of electromotive force is therefore taken 100,000,000 times greater, and is called the *Volt*. The unit of resistance is taken 1,000,000,000 times greater, and is called the *Ohm*.

If the wire of which the revolving coil and the galvanometer coil are formed is of uniform character and diameter, and if we find their joint resistance, we can easily prepare proportionate lengths of similar wire which will represent one ohm or any given number of ohms. Sets of such wires, coiled and so mounted upon an insulating support that they may be joined in any way desired, constitute what are called *resistance coils*. They are concrete standards of resistance, and in theory they are used just as standards of length, mass, etc., are used.

One of the most obvious methods of comparing the unknown resistance of any given conductor with that of a standard coil will be understood by considering the effect of friction on the flow of water in pipes. If a pipe be selected as a standard, through which, with a given constant pressure of water, exactly one gallon will flow in one second, it is evident that any other pipe which, under the same conditions, will deliver the same amount must offer the same resistance. If the pipe under trial deliver only one-tenth of a gallon, it evidently offers ten times the resistance of the standard. In the case of electrical conductors we may maintain the electromotive force constant, and determine the number of ampères which flow through the standard resistance wire, and that whose resistance is to be measured successively. Their resistances will be inversely as the number of ampères which can respectively pass through them. This would be an inconvenient method in practice, and accordingly others have been devised, but our limits forbid presenting them. [See *Wheatstone's balance* on page 30.]

The energy expended by a current in passing from one point to another of the conductor is due to its loss of potential. The equivalent work may appear in heating the conductor, as in electric lighting, in electrolytic work, or in mechanical work according to the devices included in the circuit.

The power, or, in other words, the rate at which work is done when one ampère of current suffers a loss of potential equal to one volt is called a Watt. It is 10,000,000 ergs per second, or about one seven-hundred-and-forty-sixth part of a horse-power.

An example will illustrate the use which may be made of these relations. It requires about ten ampères of current to operate an arc-

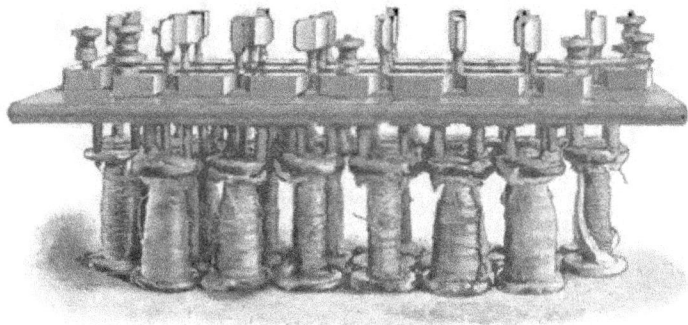

Resistance Coils, removed from box; showing their connections by means of plugs between metallic blocks.

lamp such as is employed in street-lighting. The current suffers a loss of about fifty volts at each lamp. If there be fifty lamps in the series the entire loss will be 2,500 volts. This number, multiplied by 10 and divided by 746, gives between thirty-three and thirty-four horse-power, which will be required to operate all the lamps.

In conclusion, it may be worth while to emphasize what has already, perhaps, sufficiently appeared, namely, that we can entertain no expectation that electricity of itself will ever in our hands become a *source* of energy with which we can operate factories, drive trains, etc. It can

only play a part in our service when there is a difference of potential
between the points of its application, and, in order to secure this difference of potential, work must be done, in general, greater in amount
than can be recovered from the fall of potential when equilibrium is
restored. Similar disadvantages attend the use of other means of
applying energy, such as belting, shafting, steam and water in pipes,
etc.

It would be quite impossible to forecast the future, even for a

Box of standard resistances.   Galvanoscope.   Battery.   Conductor to be measured.
Above, sliding key for effecting balance.
Wheatstone's Balance, used for comparing resistances.

single decade, with reference to the applications of electricity, even
though discovery were ended. The mere expansion of industries
already in some degree established will give them an importance which
we cannot now estimate. But discovery is not ended, and it is more
than probable that results will yet be reached which, although they
cannot be at variance with the general doctrine of energy as now understood, may, to some extent, revolutionize our methods, with corresponding advantages.

THE ELECTRIC MOTOR AND ITS APPLICATIONS.

By FRANKLIN LEONARD POPE.

FARADAY'S FIRST MOTOR—EXPERIMENTS OF AMPÈRE AND ARAGO—THE GREAT ELECTRO-MAGNET OF PROFESSOR JOSEPH HENRY—EARLY ATTEMPTS AT ELECTRO-MAGNETIC ENGINES—PROFESSOR JACOBI'S BOAT—THE ELECTROSTATIC COIL INVENTED BY DR. PAGE—HIS ELECTRIC LOCOMOTIVE OF 1851—HENRY'S PREDICTION—THE FIRST PRODUCTION OF AN ELECTRIC CURRENT BY MECHANICAL POWER—EVOLUTION OF THE DYNAMO—GRAMME'S MACHINE—REVIVAL OF INTEREST IN ELECTRIC RAIL-WAYS—SIEMENS'S LOCOMOTIVE—EXPERIMENTS OF FIELD AND EDISON—LEO DAFT'S FIRST RAILWAY IN OPERATION—OTHER APPLICATIONS OF THE MOTOR—POSSIBILITIES OF GREAT SPEED.

ON the morning of December 25, 1821, the young wife of an assistant in the laboratory of the Royal Institution of London, was called by her husband to share his delight at the success of an interesting experiment, the possibility of accomplishing which had occupied his thoughts for many weeks. What the young woman saw, upon entering the laboratory, was this: Upon a table stood a small vessel filled nearly to the brim with mercury; a copper wire was supported in a vertical position, so as to dip into the mercury, while a little bar-magnet floated in the liquid metal as a spar-buoy floats in a tideway, having been anchored by a bit of thread to the bottom of the vessel. The mass of mercury having been connected by a wire to one pole of a voltaic battery, the experimentalist had found that whenever the electric circuit was completed, by touching the other battery conductor to the vertical wire, the floating bar would revolve around the latter as a centre.   In this simple manner a continuous mechanical motion was, for the first time, produced by the action of an electric current.

The world is even now but just awakening to a conception of the magnificent possibilities of the humble gift which was slipped into its stocking on that Christmas morning by the since world-famous man.

André Marie Ampère. (After a steel engraving, by Tardieu, in 1825.)

who not long before had jocosely described himself as "Michael Faraday, late bookbinders' apprentice, now turned philosopher."

In the winter of 1819-20, the Danish philosopher Œrsted had observed that if an electric current was made to traverse a wire in proximity and parallel to a magnetic compass-needle, the needle was

Faraday Announcing His Discovery to His Wife on Christmas Morning, 1821.

3

deflected, and instead of pointing to the north, tended to place itself at right angles to the wire. The consequences of this discovery, which in truth was nothing less than that of the possibility of converting the energy of an electric current into mechanical power, proved to be far-reaching and important. It was at once seized upon by the brilliant and fertile mind of the French academician Ampère, who, by a series of masterly analyses, showed that all the observed phenomena were referable to the mutual attractions and repulsions of parallel electric currents, and with his *confrère* Arago succeeded in permanently magnetizing a common sewing-needle by surrounding it with a helically coiled wire through which an electric current was made to pass.

These brilliant discoveries inaugurated an era of active research. Faraday, as we have seen, was successful in producing continuous mechanical motion. Barlow, of Woolwich, elaborating Faraday's discovery, made in 1826 his electric spur-wheel, a most ingenious philosophical toy, and, in point of fact, the first organized electric motor. In 1826 Sturgeon devised the electro-magnet. He bent a soft iron rod into a horseshoe form, coated it with varnish and wrapped it with a single helix of bare copper bell-wire. A current passed through the wire rendered the rod magnetic, and caused it to sustain by attraction a soft iron armature of nine pounds weight.

Barlow's Spur-wheel Motor.

In this country, Professor Dana, of Yale, in his lectures on Natural Philosophy, exhibited Sturgeon's electro-magnet. Among his listeners was Morse, in whose mind was thus early planted the germ which ultimately developed into the electric telegraph. Professor Joseph Henry, then a teacher in the Albany Academy, starting with the feeble electro-magnet of Sturgeon, reconstructed and improved it, and then, by a series of brilliant original discoveries and experimental researches, developed it into an instrumentality of enormous mechanical power, capable of exhibiting a sustaining force of 2,300 pounds, a power which nevertheless vanished in the twinkling of an eye upon the breaking of the electric current.

With characteristic sagacity, the prophetic mind of Henry at once foresaw the more important uses to which his improvements were applicable. He constructed, in 1831, a telegraph in which strokes upon a bell were produced at a distance by the attractive force of the electro-magnet, which embodied all the fundamental and necessary mechanism of the electric telegraph of to-day. He also devised and constructed the first electro-magnetic motor. In a letter to Professor Silliman, in 1831, he says: " I have lately succeeded in producing motion by a little machine which I believe has never before been applied in mechanics— by magnetic attraction and repulsion." It was a crude affair and served merely to illustrate the essential principle of such an apparatus. A vibrating or reciprocating electro-magnet was provided with an attachment for controlling the current of the battery by interrupting and reversing it at the proper time. This machine, which

Professor Joseph Henry's Electro-magnetic Motor.   (From a photograph of the original.)

is of much historical interest, is now in the cabinet of Princeton College. Several large electro-magnets constructed by Henry are also among the most valued treasures of this collection.

After having thus demonstrated the possibility of constructing an operative electro-magnetic engine, so far from giving way to the natural enthusiasm of the successful inventor, Henry proceeded, with that sobriety of judgment which was perhaps his most prominent mental characteristic, to forecast the future possibilities of the new motor. He was soon led to see that in the then known state of the art the power must be derived solely from the oxidation or combustion of zinc in the voltaic battery, and hence that the heat-energy required for the original deoxidation of the metal must represent at least an equal amount of power, the inevitable corollary of which was that the fuel required for this purpose might with much more economy be employed directly in performing the required work.

Although thus well assured that electro-magnetism could never hope to compete with, much less supersede, steam as a prime motor, nevertheless he did not hesitate to predict that the electric motor was destined

Professor Joseph Henry

to occupy an extensive field of usefulness, particularly in applications to minor branches of industry in which economy of operation was subordinate to other more important considerations.

This fundamental, and as time has shown accurately, prophetic conception of the legitimate field of the electric motor, almost wholly failed to impress itself upon the minds of Henry's contemporaries. The problem of the application of electricity as a universal motive power was taken up with great zeal by a host of sanguine inventors.

In 1832, Sturgeon constructed a rotary electro-magnetic engine, of which we give an illustration, a fac-simile of his own drawing, which he exhibited before a large audience in London in the spring of 1833. In our own country, probably the earliest electric motor designed for practical use was the production of Thomas Davenport, an ingenious Vermont blacksmith, who, having seen a magnet used at the Crown Point mines in 1833 for extracting iron from pulverized ore, was seized with the idea of applying magnetism to the propulsion of machinery. In 1834 he produced a rotary electro-magnetic engine, and in the autumn of 1835 he exhibited in Springfield, Mass., a model of a circular railway which was traversed by an electro-magnetic locomotive.

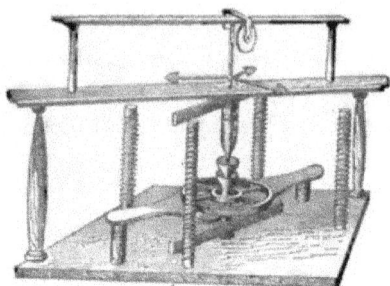

Sturgeon's Electro-magnetic Engine.

Many citizens of New York will recall the erect and handsome figure of a venerable gentleman, with flowing white hair, dressed with scrupulous neatness in the Continental costume and cocked hat of the period of the Revolution, who was to be seen on Broadway every pleasant day early in the seventies, and whose resemblance to the accepted portraits of Washington was so striking as to at once arrest the attention of the observer. This was Frederick Coombs, who, as the agent of Davenport, visited London, in 1838, where he exhibited a locomotive weighing 60 or 70 pounds, propelled around a circular railway track by electric power, which excited the greatest interest in the scientific circles of the metropolis.

In 1840, Davenport printed by the power of an electric motor a sheet entitled the *Electro-Magnet and Mechanics' Intelligencer*. Meantime others had occupied themselves with similar undertakings. Professor Jacobi, of St. Petersburg, invented a rotary electro-magnetic motor in **1834**, and with the financial assistance of the Emperor Nicholas constructed, in 1839, a boat 28 feet long, carrying 14 passengers, which was propelled by an electric motor, with a large number of battery cells, at a speed of 3 miles per hour. In 1838–39, Robert Davidson, a Scotchman, experimented with an electric railway car 16 feet long and weighing, including the batteries, 6 tons, which attained a speed of 4 miles per hour.

The necessary limitations of space preclude even the briefest notice of the labors of the host of ingenious experimenters who occupied themselves in this attractive line of scientific research; but no historical sketch of the electric motor would be complete without some reference to the work of the late Dr. Charles Grafton Page, who for many years occupied a responsible official position in the Patent Office at Washington.

Page, while a young medical student in Salem, Mass., entered upon an experimental investigation of the relations between electricity and magnetism, and he continued to prosecute his researches with extraordinary diligence and success during the greater portion of his active life. He particularly

Dr. Charles Grafton Page

distinguished himself by his researches in electrical induction, notably by his invention of the electrostatic induction coil and circuit-breaker, which has been persistently, but without a shadow of justice, attributed to Ruhmkorff, of Paris. His work in connection with the electric motor, although not so well known, owing to the scanty information which is contained in contemporaneous records, is certainly no less important. Many middle-aged men of to-day will recall the interesting and curious array of apparatus for illustrating electro-magnetic rotation which formed such an important part of the philosophical cabinets of the academies and colleges of the preceding generation. It is not too much to say that almost every one

of these devices owes its origin to the fertile and ingenious brain of Page.

As early as 1845 it had been observed by Alfred Vail, the coadjutor of Professor Morse in the construction of the electric telegraph, that a hollow coil of insulated wire, when traversed by an electric current, possessed the curious property of sucking a soft iron core into itself with considerable force. Upon this phenomenon being shown to Dr. Page, he at once conceived the idea of utilizing it in the operation of an electric motor, and after numerous experiments he succeeded in constructing, in 1850, a machine of this description, which developed over 10 horse-power. Aided by an appropriation from Congress, he subsequently constructed an electric locomotive of considerable size with which an experimental trip was made from Washington to Bladensburg, on the Washington branch of the Baltimore and Ohio Railroad, on April 29, 1851, and on which occasion a rate of speed was attained, on a nearly level plane, of 19 miles per hour. Of course in this, as in other experiments which have been detailed, the great cost of producing electricity by the consumption of zinc in a battery precluded the possibility of any commercial advantage being derived from the scheme, but the achievement was nevertheless a notable one. Not far from the same time Thomas Hall, of Boston, a skilful electromechanic who had constructed much of Page's apparatus, made a small model of an electric locomotive and car, which is not without scientific interest, as establishing the practicability of conveying the electric current to a rapidly moving railway car by employing the rails and wheels as electrical conductors, thus dispensing with the necessity of transporting the battery upon the vehicle.

One of the most enthusiastic experimentalists with electro-magnetic machinery was the late Dr. James Prescott Joule, of Manchester, England, who in a letter written in 1839, said: "I can scarcely doubt that electro-magnetism will eventually be substituted for steam in propelling machinery." Professor Jacobi, too, one of the most eminent philosophers of that day, was less reserved in his enthusiasm than

Henry, for he wrote: "I think I may assert that the superiority of this new mover is placed beyond a doubt as regards the absence of all danger, the simplicity of action, and *the expense attending it.*"

Some years afterward, when Dr. Joule had become older and possibly wiser, he made a most remarkable and exhaustive series of investigations relating to the mechanical equivalent of heat. The results of these researches led him to estimate that the consumption of one grain of zinc can produce only about one-eighth of the mechanical equivalent of a grain of coal, while its cost is approximately twenty times as great. This conclusion, being generally accepted by the scientific world as authoritative, ultimately tended to strongly discourage further efforts to apply electro-magnetism as a prime motor. The question was well summed up, in a discussion by Henry, in these words: "All attempts to substitute electricity or magnetism for *coal-power* must be unsuccessful, since these powers tend to an equilibrium, from which they can only be disturbed by the application of another power, which is the equivalent of that which they can subsequently exhibit. They are, however, with chemical attraction, etc., of great importance *as intermediate agents in the application of the power of heat as derived from combustion.* Science does not indicate in the slightest degree the possibility of the discovery of a new primary power comparable with that of combustion as exhibited in the burning of coal. . . . We therefore do not hesitate to say that all declarations of the discovery of a new power which is to supersede the use of coal as a motive power have their origin in ignorance or deception, and frequently in both."

In the words which have been italicized Henry accurately foretold the true place, in the domain of industry, of the electric motor. Much confusion of thought exists in the popular mind at the present time in reference to this very point. We continually hear electricity spoken of as if it were a *primary, a motive power,* and the prediction is freely made by uninformed persons that it will soon usurp the place of the steam-engine; that it will be employed to propel vessels across the Atlantic, and the like. But a moment's consideration will serve to show that such a view of the question is wholly without scientific basis. Electricity, in its important applications to machinery, is never in itself a source of power. It is merely a convenient and easily manageable agency, perhaps a peculiar form of matter, perhaps a form of energy, by which mechanical power may be conveniently transferred from an ordinary prime motor, as a steam engine or a water-wheel, to a secondary motor—it may be at a great distance—which is employed to do the work. It performs an office precisely analogous to that of a belt or line of shafting. which, however useful in conveying mechanical power from one point to another, can under no conceivable circumstances be capable of originating it.

That we may properly comprehend and appreciate this new and important aspect of the mechanical application of electricity, it is necessary to return to the experiments of Faraday. In 1831, after he had become the director of the laboratory of the Royal Institution, he turned his attention to what he called the "evolution of electricity from magnetism." The brilliant generalizations of Ampère, followed by the experimental demonstrations of Arago, Sturgeon, and Henry, to the penetrating mind of Faraday necessarily implied *reciprocal* *action*, and he accordingly sought diligently to obtain the electric current from the magnet. On the second day after commencing his experiments he wrote to a friend: "I think I have got hold of a good thing, but cannot say; it may be a weed instead of a fish which after all my labor I may pull up." On the tenth day he became fully satisfied that he had hooked a fish. A crucial experiment showed that he had made a grand discovery which may. without injustice, fairly be compared, in point of practical importance, with Newton's immortal discovery of gravitation. The principle upon which this discovery hinges may be explained in a few words. Every magnet is surrounded by a sphere of attraction (which gradually diminishes in intensity as the distance from the pole or focal point of attraction of the magnet increases) which has received the technical name of the *magnetic field*. If an electric conductor be moved transversely through this magnetic field, the influence of the field tends to retard or oppose the movement of the conductor; the mechanical force exerted in overcoming this resistance is transformed and appears in the conductor in the form of what we conventionally call an electric current. If instead of the magnet we substitute another wire, which is

Arago.

also conveying an electric current, this last is surrounded by a magnetic field, and similar phenomena are manifested when another wire is made to move within it. The same manifestations occur if the conductor remains stationary and the magnetic field is moved, or if the strength of the field be increased or diminished. This effect is known by the general name of *induction*, and the law which governs it was formulated by the Russian philosopher Lenz as long ago as 1833. It may be stated as follows: The currents induced by the relative movements either of two electric circuits, or of a circuit and a magnet, are always in such directions as to produce mechanical forces tending to stop the motion which produces them.

To Faraday is due the first experimental machine for the mechanical production of electric currents. But he went no further. He possessed preëminently the scientific mind. His pleasure in the pursuit of natural truths was so absorbing that he could never turn away from them for the mere purpose of following up their practical applications. " I have rather been desirous," he once wrote, " of discovering new facts and new relations dependent on magneto-induction, than of exalting the force of those already obtained, being sure that the latter would find their full development hereafter." In the words of Professor Sylvanus Thompson : " Can any passage be found in the whole range of science more profoundly prophetic or more characteristically philosophic than these words with which Faraday closed this section of his Experimental Researches ? "

Within a year after the publication of Faraday's experiment, Pixii, a philosophical instrument maker of Paris, constructed an apparatus in which a permanent magnet was made to induce currents in the wire surrounding an electro-magnet ; this received the name of the magneto-electric machine, and was doubtless the first organized appliance for producing an electric current by mechanical power. In 1838 this instrument was materially improved by Saxton, of Philadelphia, whose apparatus will be recognized as the well-known " electric shock machine," in which electric currents are produced by turning a crank. A similar device is utilized at the present day for ringing telephone call-bells. For many years the practical applications of the magneto-electric machine were comparatively unimportant, and were principally confined to its employment for actuating certain forms of telegraph apparatus, thereby dispensing with the voltaic battery.

In 1850 Professor Nollet, of Brussels, essayed to make a powerful magneto-electric machine for decomposing water into its constituent

elements, oxygen and hydrogen, which were then to be used in producing the lime-light.   In 1858 a company was organized in Paris,
and experiments were made with a large machine constructed by
Nollet. So far as the lime-light scheme was concerned the experiments were unsuccessful, but subsequently Mr. F. H. Holmes, of
England, made some alterations in Nollet's machine, and applied
it directly to the production of the electric light between carbon points.
These experiments induced others to take up the subject both in
France and England, which ultimately resulted in the development of
the brilliant and beautiful electric arc-light, by which the streets of
our principal towns are now nightly illuminated.   It has been used in
some of the French lighthouses without intermission since 1863.

The substitution of the electro for the permanent magnet, first suggested by Wheatstone in 1845, was applied in the construction of large
machines by Wilde, of Manchester, who worked at the subject continuously from 1863 to 1867, with results incomparably in advance of all
previous attempts to obtain electricity by mechanical power.   In 1867
he exhibited a machine which produced the electric arc-light in its
utmost magnificence, and was capable of instantly fusing iron rods
fifteen inches long and one-fourth of an inch in diameter by the flow of
the electric current.

The final step in the development of the magneto-electric generator
was an almost simultaneous, although independent, discovery by Moses
G. Farmer, of Salem, Mass., Alfred Varley and Professor Charles
Wheatstone, of England, and Dr. Werner Siemens, of Berlin.   This
was the grand conception of employing the current from an electro-
magnetic machine to excite its own electro-magnet.   The invention of
this improved form of apparatus, which received the name of the
dynamo-electric machine, gave an extraordinary impetus to the investigation of all branches of electric science.   The subject was once
more taken up by scores of enthusiastic workers in Europe and America, and innumerable minor improvements rapidly succeeded one another, which have finally resulted in the exquisitely organized dynamo

machine of to-day, a machine which has confessedly reached a state of perfection and efficiency which leaves but the narrowest margin for future improvement.

As we have seen, the earliest field of usefulness for the dynamo-machine was found in electric arc-lighting, which has now become, in the United States at least, an enormous industry. One of the most useful and convenient of these machines was designed by Gramme, of Paris, in 1872, and was capable of giving a constant current resembling in its characteristics that from a battery. At an industrial exhibition in Vienna, in 1873, a number of Gramme machines were being placed in position, in order to exemplify the various uses to which the invention might be put as an electric generator, when there occurred one of those singularly fortunate accidents which have again and again played so prominent a part in the history of industrial progress. In making the electrical connections to one of these machines which had not as yet been belted to the engine-shaft, a careless workman attached to it by mistake a pair of wires which were already connected with another dynamo-machine which was in rapid motion. To the amazement of this worthy artisan the second machine commenced to revolve with great rapidity in a reverse direction. Upon the attention of M. Gramme being directed to this phenomenon, he at once perceived that the second machine was performing the function of a motor, and that what was taking place was an actual transference of mechanical power through the medium of electricity. This singularly opportune occurrence being commented upon in the scientific journals, led to the instant recognition of the true place of the electric motor in the domain of mechanics. From the date of Page's experiments almost the only practical use to which the electric motor had been applied was in the operation of dental apparatus, to which it has been adapted with great ingenuity and success by some of the American practitioners of that art.

The late Professor James Clerk Maxwell, one of the master minds among the electricians of the new era, shortly before his lamented death, expressed the opinion that the reversibility of the Gramme machine was one of the most important discoveries of modern times. While it is true that the circumstance attracted general attention in scientific circles, its application to useful purposes was no doubt deferred for many years by the counter attraction of electric lighting, which seemed to promise to inventors and capitalists larger and more immediate profits. The principle of converting mechanical energy into electric currents and again re-converting these by means of a reversed dynamo-machine into mechanical power, naturally suggested the practicability of transmitting power through electric conductors to any required distance. One of the earliest applications of this character was the revival of the electrically operated railway. This, as we have seen, was, abstractly considered, by no means a novel idea, but its commercial development for obvious reasons remained in abeyance until generating machinery was available which could be made to furnish large quantities of electricity at moderate cost.

One fact of controlling importance in this connection is that electricity is capable of being supplied to a moving motor through frictional or rolling contact, a method of communicating mechanical power impossible of realization by other known means; hence the power could readily be supplied by machinery situated at any required distance from the moving train by extending the conductor the required distance along the railway, for which, as we have seen, the rails themselves, when properly insulated, might be made to serve. It is probable that the earliest detailed conception of the modern electric railway was due to Jean Henry Cazal, a French engineer, who proposed, as early as 1864, to utilize the natural powers, such as water and wind, for operating railways, by the electrical transmission of power. But this was in the day of small things in electric generators, and hence the practical realization of the ideas of Cazal was only rendered possible at a much later date by the subsequent development of the dynamo-machine, as heretofore related.

A request by the proprietors of a German colliery to be supplied with an electric locomotive for hauling coal cars in the levels led Dr. Werner Siemens to devise and construct an electric railway, which was exhibited at the Industrial Exhibition in Berlin, in the summer of 1879. This railway was circular, about 1.000 feet in length, and of one metre gauge. A dynamo-electric machine driven by a steam-engine supplied

Dr. Werner Siemens.

the current, the expenditure of energy being about five horse-power. More than 100,000 persons are said to have been transported over this line during the period of the exhibition.

Meantime several American inventors were independently at work upon this problem. among them Stephen D. Field, of San Francisco: Dr. Joseph R. Finney. of Pittsburgh. and Thomas A. Edison, of New York. Edison was the first to construct an actual dynamo-electric railway in America. This was done in the spring of 1880, at the expense of Henry Villard, at Menlo Park, N. J., the track being some 80 or 90 rods in length. Field's electric locomotive was first exhibited at the Exposition of Railway Appliances in Chicago, in June, 1883, during the continuance of which nearly 27,000 passengers were transported. Both Field and Edison utilized the rails of the track to convey the current to the motor.

Finney's plan was somewhat different. He suspended an insulated copper wire, about the thickness of a lead-pencil, 15 or 20 feet above the line of the railway. A small wheeled trolley or contact-truck running on this wire as on a track, and connected with the car by a flexible conducting cord, served to convey the electric current from the aërial conductor to the motor; a plan which in practice was found to answer very satisfactorily for moderate speeds. Finney designed from the first to apply the electric motor to an ordinary street-car as a substitute for horse-power. His first experimental car was exhibited and very successfully operated in Allegheny, Pa., in the summer of 1882.

The first electric street railway operated in America for actual commercial service was a suburban line two miles in length extending from Baltimore to Hampden, Md. It had previously been operated by animal power, and was cheaply and roughly constructed, having sharp curves, and grades as high as 330 feet per mile. This line was put in operation September 1, 1885. The electric current was conveyed by an insulated rail fixed to the ties midway between the traffic rails. The electrical machinery was designed and constructed by Leo Daft, of Jersey City, N. J. The results of the change of motive power were gratifying, inasmuch as the receipts of the line were largely increased during the first year, while on the other hand, the expense of operation was diminished, and this in spite of the fact that the application was made under exceptionally unfavorable circumstances. The success of this undertaking went far to demonstrate the advantages of electricity as a street-car motor.

Every consideration of humanity, no less than of convenience and

4

economy, unites to urge the substitution of mechanical for animal power
upon the numerous street railway lines of the United States at the
earliest practicable moment,* and hence it is gratifying to know that
on the first of August, 1890, there were in daily service, and under
construction, in the United States and Canada, no less than 250 street
railways operated by electricity, having a total length of between 1,900
and 2,000 miles, while a very large number of others are under contract. On January 1, 1888, there were but 23 of these lines, having a
total length of about 100 miles.

One of the earliest successful examples of an electric street railway
was constructed at
Scranton, Pa. It was
designed by Charles J.
Van Depocle, then of
Chicago, and went
into operation in Decem-
ber, 1886. It was four
and one-half miles in
length, of standard
gauge, laid with steel
rails, and its passenger
equipment consisted of
seven handsomely fin-
ished Pullman cars, each

The Van Depoele Electric Motor.

propelled by a 15 horse-power electric motor, which was placed on
a glass-enclosed front platform and geared to the forward axle by
the familiar mechanical device of sprocket wheels and steel chains.
The external appearance of the motor used on the Scranton cars is
shown in the illustration. It stands about two feet high, and occupies

* In the street railway service in large cities, the distance travelled each day by a
two-horse team averages about ten miles, so that each animal works only about two
hours out of the twenty-four. The cost of stabling, feeding, and replacing horses is
$200 per year, each. The active life of a car-horse is only from two to four years.

a space perhaps eighteen inches square. The car, of which an illustration is given, could be run at a speed of fifteen miles per hour, if
required, and in its regular work ascended grades of nearly 350 feet
per mile with great facility. It is stated
that the cost of running the electric line at
Scranton, using for fuel the waste coal-dust
or "culm" from the anthracite mines, which
can be had in almost inexhaustible quantity
at the nominal price of 10 cents per ton, was

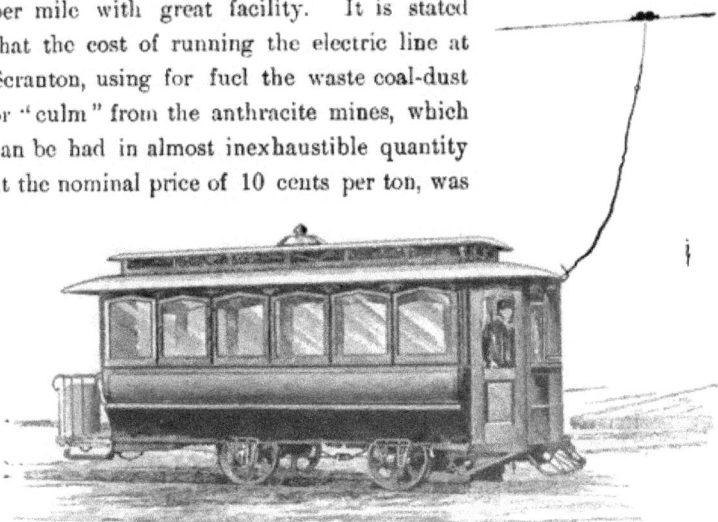

A Street-car Propelled by an Electric Motor.

about one dollar per car per day, or a trifle over one cent per car mile.
The economy even at this early stage of development of electric power
over animal power, the cost of which in New York and Boston is
reckoned at something over ten cents per car mile, is apparent.

Electric railways are in operation at Appleton, Wis., St. Catharine,
Ontario, and other places, which are driven by water power at an
almost nominal cost. In many instances natural power may be thus
used with the utmost advantage, as it is not in the least necessary that
the power should be in the vicinity of the line of the railway.

Several of the inventors whose names have been already mentioned
in connection with electric railway work have paid much attention to
the problem of city and suburban rapid transit, and there is every
reason to hope that an early day will witness the successful introduc-

tion of electric power upon the elevated railway system in New York. for which it would seem on many accounts to be peculiarly well adapted.

The pioneer electric street railway in Europe was the Lichterfelde line, in the suburbs of Berlin, constructed under the superintendence of Dr. Siemens, which has been running since the spring of 1881. Several other electric lines have been constructed in Great Britain, Ireland, and on the Continent, but, as usual with inventions of this class, our own country quickly placed itself far in advance of all others in the extent to which the new system found a practical application.

A considerable number of street-cars have been constructed both in Europe and America, with the design of deriving the electric energy for propelling each individual car from storage-batteries or accumulators carried upon it.* An accumulator may be described, in a general way, as a vessel containing acidulated water in which is immersed a pair of leaden plates. The passage of a strong electric current from a dynamo, through the liquid from one plate to the other, produces a chemical action which has the effect of oxidizing one of the plates. After this process has gone on for some hours, the dynamo may be detached, and the two plates joined by a wire. An electric current will now pass through the wire from one plate to the other, as in an ordinary voltaic battery, the effect of which is to undo the work which has been done in charging the battery. Strictly speaking, in a battery of this kind no electricity is stored; its energy is in fact converted into chemical energy, and this may be reconverted into electric energy at will. Many quite successful experiments have been made with self-propelling motor cars operated by these batteries, and there is little reason to doubt that they will ultimately find an extensive use and application in large cities and other localities where the employment of an overhead conductor for electrical distribution is from any cause objectionable. The most serious objection to their use is the

---

* See "The Electric Railway of To-day."

expense of operation, **which has thus far** proved **to be** much **greater than** in the case of direct supply from **the dynamo.**

Many inventors are endeavoring to find a thoroughly practicable plan of insulating the electric conductor beneath the roadway; **and** while the problem is unquestionably a far more difficult one than would be supposed by a person unfamiliar with the subject, yet there is probably no good reason to doubt that it will in due time receive a practical solution.

The beginning of the general introduction of electric lighting by incandescent lamps supplied from central stations, which may be fairly considered to date from about 1883, had the almost immediate effect of creating a demand for small electric motors. It was at once perceived that the electric lighting conductors, if introduced into every building in a town and supplying a constant electric current, at an expense ordinarily not exceeding 8 or 10 cents per horse-power per hour, could be utilized with great advantage in driving sewing-machines, lathes, ventilating apparatus, and innumerable other sorts of machinery for domestic purposes, or for the lighter class of mechanical industries. Quite an assortment of neat little motors of this character, of different patterns, and of capacities ranging from one-tenth to one-half a horse-power (for operating sewing-machines, and other light work), were exhibited in 1884 at the Industrial Electrical Exhibition at Philadelphia, where they attracted much attention.

One of the most interesting exhibits of this character was made by Lieut. F. G. Sprague, formerly an officer in the United States Navy, who showed two or three motors of his own design having a capacity of perhaps five horse-power, which were employed to drive looms and other textile machinery requiring considerable power. Another motor of about two horse-power, built by Mr. Daft, was at work for several weeks during the exhibition, printing the regular weekly issue of an electrical journal, on a power press with a bed 31 by 46 inches. The successful and satisfactory operation of these motors led almost immediately to the establishment of an extensive business, and there are now in New York, Boston, Troy, Rochester, and other cities, systems of electric power-distribution from central stations of considerable importance, employing

The Sprague Motor Running an Elevator.

machines of the general type first exhibited at Philadelphia on the occasion just referred to.

It is a very difficult matter to ascertain even approximately the extent to which this business of electric power-distribution has already attained in this country, but a somewhat cursory investigation has shown that it is greatly in excess of what might have been anticipated. One central power station in Boston operates more than one hundred motors of a capacity ranging from 15 down to one-half a horse-power, the greater number used being from 5 to 10 horse-power. The supply conductors are carried underneath the pavement of the streets. A single corporation of the dozen or more actively engaged in this manufacture has sold within five years over 4,000 motors, aggregating more than 20,000 horse-power, and the demand is increasing daily. It would be almost impossible to catalogue the number and variety of purposes for which the electric motor is now in daily use. Some of the most usual applications are for printing-presses, sewing-machines, elevators, ventilating-fans, and machinist's lathes.* At the present time every indication unmistakably points to the probability that within a very few years nearly all mechanical work in large cities, especially in cases in which the power required does not exceed say 50 horse-power, will be performed by the agency of the electric motor. It is an ideal motor, absolutely free from vibration or noise, perfectly manageable, entirely safe, and with the most ordinary care seldom if ever gets out of order. Indeed, there is no reason to suppose that the limit of 50 horse-power will not be very largely exceeded, when it is remembered that scarcely five years ago the production of a 10 horse-power motor was considered quite a noteworthy achievement, while at the present day motors of 300 horse-power, and even more, have been successfully installed.

Motor attached to a Sewing-machine.

An extremely useful application of the electric motor, which is likely to be widely extended, is in connection with large manufacturing establishments, already supplied with incandescent electric lighting apparatus. It is a very simple matter, by means of a current derived from the same dynamo, to operate elevators, hoists, presses, pumps, trucks, tramway-cars, and many similar appliances, which are now

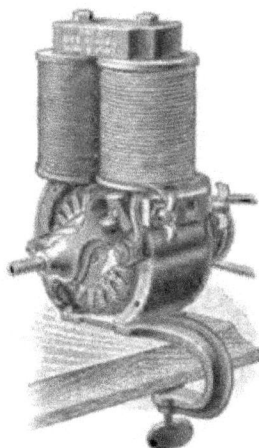

* See "Electricity in the Household."

worked at greater expense, and with far less convenience, by hand, animal, or independent steam power. We give an illustration of one of the earliest successful examples of this description of tramway work at a sugar refinery in East Boston, Mass. The electric motor is geared to the axle of a low platform car, and serves to propel it, together with a second car, along the track, the whole being operated by a current from the incandescent dynamo used for lighting the premises. This little freight train makes a round trip every five minutes under the management of an ordinary laborer, hauling an average

Electric Tramway for Hauling Freight, used in a Sugar Refinery.

load of ten tons of raw sugar from the wharf to the refinery at each trip, at an inconsiderable expense.

Another very important service to which the electric motor is especially well adapted is that of a substitute for belts, shafting, and gearing, in the transmission of power from the prime motor in large manufacturing establishments. A New England cotton-mill engineer of high repute has ascertained, from actual measurement of a number of modern mills fitted with first-class shafting, that over thirty per cent. of the gross power of the engines is absorbed in driving the various lines of shafting alone, before the delivery of any power whatever for actual work. Numerous tests demonstrate that it is entirely within the truth to estimate the loss of conversion, transmission, and reconversion, in well-designed electrical machinery, under like conditions, at less than thirty per cent., so that in the use of the electric motor for this class of work, we have at once an actual saving over the loss experienced in the direct mechanical transmission of power, with the further and in most cases controlling consideration, that in the case of the electrical system this loss affects only such portions of the machinery as are actually at work; while under the ordinary conditions the entire system of belting and shafting must be kept in continuous operation, entailing a constant loss, irrespective of the number of machines which may be actually in use at any given time. The advantages of having every individual machine driven by its own independently controlled power, and at any required speed, are so obvious that it is scarcely necessary to mention them.

The conditions of electrical power transmission have been thoroughly studied by competent engineers, and are now so well understood that those conversant with the practical aspects of the subject are well assured that within a few years even the smallest towns and villages will supply themselves with electric light and power plants. In such places a plant of 50 horse-power, or even less, will be quite sufficient to furnish a good profit on the moderate investment of capital required. The establishment of a power centre, even in a rural village, cannot fail to attract a greater or less number of small though by no means unprofitable industrial enterprises, and the mere fact that such power can be had will in itself tend to rapidly increase the demand. The management of an electric power plant requires no unusual scientific knowledge. Once the station has been established, it can be carried on by the ordinarily intelligent class of mechanics and workmen who are to be found in every village. It is computed by statisticians that the average price at which power is sold in the United States approximates $110 per horse-power per annum. A 50 horse-power electrical plant, including the station building, engines, boilers, dynamos, distributing wires, and fixtures, can be erected, at present prices, at an expense not much exceeding $150 per horse-power, and the gross cost of operating such a plant may be fairly estimated at about $4,000 per year. Experience has shown that in consequence of the intermittent demand for power by a group of miscellaneous consumers, it is entirely safe to contract to supply a quantity considerably in excess of the actual capacity of the station, so that indeed as much as 70 horse-power might be sold from a 50 horse-power plant, thus bringing in a yearly gross revenue of $7,000 or more, and leaving a net profit of some $3,000. Where a good water-power is available at a moderate outlay, the profits might be even more than we have estimated, while it will be readily understood that in all such cases the proportionate profits are rapidly augmented as the capacity of the plant is increased.

A somewhat startling proposition in connection with the general subject of the transmission of energy to a distance by electricity was advanced by that eminent engineer, the late Charles W. Siemens, of London, who, in 1877, expressed his conviction that by this means the enormous energy of the falling water at Niagara might be transferred to New York City, and there utilized for mechanical purposes. In 1879, Sir William Thomson, the electrician, publicly asserted his belief in the possibility, by means of an insulated copper wire, half an inch in diameter, of taking 26,000 horse-power from water-wheels driven by the falls, and of delivering 21,000 horse-power at a distance of 300 statute miles. He estimated that the cost of copper for the line would be less than $15 per horse-power of energy actually delivered at the remote station. While Sir William may be regarded as somewhat of an enthusiast, and has occasionally manifested a tendency to present matters of this kind in a sensational light, yet it cannot be looked upon as especially improbable that the realization of this apparently chimerical project will be witnessed by persons now living. In fact, engineering works of the most extensive character, for the utilization of the power of Niagara for electrical distribution, are already actually under way.

Space does not permit more than a passing reference to the important industrial applications of the electric motor for the transmission of power which have been made within the past two or three years, both in Europe and America. At Solothurn, Switzerland, a screw factory has been operated for three years past by a double 50 horse-power motor, deriving its energy from a water-fall 5 miles distant ; and at Derendingen, in the same vicinity, a manufactory of white woollen goods of 36,000 spindles is driven entirely by a pair of electric motors of 280 horse-power. At Schaffhausen, an electric motor of 400 horse-power, deriving its energy from the famous Rhine fall at that place, has been put in successful operation.

We have as yet no examples of single motors of such large capacity in this country ; the tendency here has been rather in the direction of the distribution of power from a common centre among a number of small industries, a mode of application which, without doubt, affords more remunerative commercial returns than the transmission of power in large units, although perhaps less striking to the ordinary observer as an example of the possibilities of the system. At Aspen, Col., a water-power electric plant has been installed, consisting of eight 24-inch Pelton wheels under a head of 820 feet. These wheels make 1,000 revolutions per minute and develop in the aggregate 1,400 horse-power. The electric plant comprises six motors of 20 horse-power and one of 60 horse-power, operating mining machinery, stamp-mills, etc., at various places within a radius of one to two miles. There are also 120 arc-lights and 2,000 sixteen candle-power incandescent electric lights which are supplied with electric current from the same source. Another interesting example of the same class of work is at the Chollar shaft on the celebrated Comstock lode in Nevada. Here are six Pelton wheels at the bottom of a shaft nearly one-third of a mile beneath the surface of the ground, running at 900 revolutions per minute, under the enormous head of 1,680 feet. The aggregate horse-power developed is 750, which is conveyed by wire a distance of one mile and employed to operate an extensive stamp-mill. At Big Bend, on the Feather River, in California, a power plant was used for some time, in which the electric circuit was carried along the river for 18 miles, operating 14 independent electric motors, located at different points and used with great facility in pumping, hoisting, hauling, and other mining operations.

A most interesting example, among many others that might be mentioned, of the utilization of electric power in small industries, is to be found in the villages of Johnstown and Gloversville, N. Y., well known as the principal seat of the glove-manufacturing industry in this country. Here Cayadutta Creek furnishes a trustworthy water power of 500 horses in the dryest season. Seven dynamos furnish electric power to six electric circuits, 3 of which extend to Johnstown and 3 to Gloversville, the aggregate length of the whole being 60 miles. Nearly 300 horse-power of electric motors are now in use in these towns of a capacity varying from one-eighth to 10 horse-power, which are used not merely for operating glove-sewing machines, but for almost every conceivable kind of village industries. Each sewing-machine operator is furnished with power at the merely nominal price of 25 cents per week, a sum which nevertheless has been found to furnish a handsome profit to the electric company. Similar works have been carried out in Rochester, N. Y., and other places, while the projects of a like nature under consideration in other localities are too numerous even for the most casual reference.

The experiment of Jacobi, who was the first to propel a vessel by electricity, has already been noticed. The records of electric navigation are a blank from that time until the commencement of the experiments of the ingenious and versatile Trouvé, of Paris, who exhibited a small boat on the Seine in 1881, the electricity for which was supplied by a primary battery. Four or five passengers were carried at a speed of about three miles per hour. Between 1882 and 1886 quite a number of experimental launches were built in England and France, propelled by electric motors, and supplied with electricity by accumulators stowed in the bottom of the boat, which served also as ballast. The most noteworthy achievement of this character was the launch Volta, which, in September, 1886, performed the trip from Dover to Calais and back, with ease and safety, the batteries being charged but once for the double journey. Seven passengers were carried on this occasion, and a speed of over twelve miles per hour was reached. The Volta was 37 feet long, seven feet beam and three and a half feet deep.

Quite recently a New York establishment which manufactures small motors, one of which is shown in the illustration on page 55 as applied to a sewing-machine and deriving its power from a primary battery, has adapted a similar motor to the propulsion of a light canoe. The dimensions of the motor designed for this craft are so small that at first sight it seems almost absurdly inadequate for its intended purpose. But its performance is nevertheless excellent, and a considerable demand has sprung up for these neat and ingenious little vessels.

It would be impossible, within the limited space at command, to attempt to enumerate the various future applications of the electric motor which suggest themselves to the enterprising electro-mechanician; but, in conclusion, the writer cannot refrain from expressing his conviction that the day is not far distant when rapid transit between the principal cities of America will be effected to an extent which to persons unfamiliar with the developments of electricity must seem utterly visionary and chimerical. Once admit, as we must do, the possibility of applying almost limitless electric power to each axle of a train, with the possibility of laying a track almost as straight as the crow flies from city to city, rising and falling as the topography of the country may require, and the complete solution of the problem becomes little more than a matter of detail. Not that such detail is unimportant, nor that the innumerable minor difficulties can be overcome without much experiment and study; but it may nevertheless safely be affirmed that the ultimate result is already distinctly foreshadowed, and that we may expect within a few years to be transported between New York and Boston in less than two hours, not by the enchanted carpet of the Arabian Nights, but by the potent agency of the modern electric motor.

THE ELECTRIC RAILWAY OF TO-DAY.

By JOSEPH WETZLER, M.E.

Early Attempts at Electric Motors—Why They were Failures—Importance of the Continuous-Current Dynamo—The Berlin Exposition Railway of 1879—Three Methods of Applying the Current to the Motor—Construction of the Truck—The Overhead-Wire System—How the Trolley is Attached—Methods of Suspending the Wires—Single and Double Poles—Sliding Contact—The Underground Conduit System—The Storage-Battery System—Advantages of Electric Street Railways over all Other Plans—Comparative Cost of Construction—Statistics for the United States—Electric Railways in Mines—Telpherage—High Speed Experiments—Future Possibilities.

IT has been remarked truthfully that the civilization of a country may be gauged by its methods and means of communication, and the transition from the stage-coach of old to the lightning express of to-day marks as great advance in the methods of passenger transportation, probably, as does that of the telegram over the post message. But while these improvements in methods of highway transportation have been going on steadily for over fifty years, with the brilliant results well known to all, there is one class of traffic which, even up to within a short time, has remained perfectly stationary since its inception; and that is, the street-car or tramway traffic. Beginning with the horse as the motive power, over fifty years have passed without an essential change in the method of propulsion, and it has remained for that subtile and vigorous agent, electricity, to solve the problem which has taxed the capacity of engineers for half a century. Attempts, it is true, have been made to displace the horse by mechanical power, applied in the shape of the steam and compressed-air locomotives, and again by the more recent cable; but the objections to their employment in the crowded streets, together with the now acknowledged superiority of the electric railway, allow of the assertion being safely made that, except in very rare cases, the former must now be considered methods of the past, and that the long serfdom of the horse will be brought to an end by the electric motor applied to the street-car.

As brilliant an achievement as the electric railway of to-day undoubtedly is, it has had its period of development, like every other modern industrial application of importance; and the period from its inception to final consummation was indeed by no means a short one. The reasons for this are, however, traceable to the same causes which so long retarded the introduction of the electric light; the long delay being due to the absence of a sufficiently powerful and economical generator of electricity. To the student, the tracing of the history of this development presents a most interesting line of study and research, but the limits of this chapter forbid our entering upon it except to briefly mention the early workers in this field.

As far back as 1835, Stratingh and Becker, of Groeningen, and Botto, of Turin, in 1836, constructed crude electric carriages. They were shortly followed by Davidson, a Scotchman, who, in 1838–39, built an electric car weighing five tons, with which he obtained a speed of four miles an hour. These were contemporaneous with others in the United States, where Thomas Davenport, a blacksmith of Brandon, Vt., built a small circular railway at Springfield, Mass., in 1835, which he operated by means of electricity. It is also worthy of note here that to Davenport, probably, belongs the honor of having first printed a newspaper by electricity, one called the *Electro-Magnet and Mechanics' Intelligencer*, in 1840. Foremost in the ranks of American pioneers in this field, however, was Professor Page, of the Smithsonian Institution, some account of whose works is given in another chapter.*

* See "The Electric Motor and its Applications."

The railroad experiments of this scientist consisted in the operation of an electric locomotive between Washington and Baltimore, in the course of which he obtained on one occasion a speed of nineteen miles an hour; but the difficulties experienced with the Grove primary batteries on the car were such as to force him to abandon the scheme. The work, in this field, of Professor Moses G. Farmer, in 1847, and of Thomas Hall, who exhibited a model electric locomotive at the Charitable Mechanics' Fair in Boston, in 1851, can only be mentioned.

All these experiments, however, interesting as they were from a scientific standpoint, were destined to practical failure on account of the enforced employment of batteries as the source of electrical energy; and it was not until the invention of the continuous-current dynamo-electric machine that the actual solution of the problem became possible. Soon after the invention of the dynamo, Siemens and Halske, of Berlin, made some attempts to apply electricity to railroad purposes; but the imperfections of the early machines led to the abandonment of the project.

But the advances which had been made in the art of dynamo-building, and the discovery of the reversibility of the dynamo, so that it could be employed as a motor, led to renewed attempts, and at the Berlin Exposition of 1879, this same firm operated a small electric railway, which was perhaps the first commercial electric railway in the world opened for regular traffic. American inventors, however, had by no means been idle, since almost at the same time Stephen D. Field, the nephew of Cyrus W. Field, of Atlantic cable fame, and Thomas A. Edison, had conceived the idea of the modern method of operating electric railways; and it is interesting to recall these attempts, as showing the lines on which these early experiments were carried out. This is illustrated by the locomotive constructed by Mr. Edison at Menlo Park, in 1880, shown in the engraving, Fig. 1, after a photograph preserved in Mr. Edison's library.

These experiments encouraged other inventors in this country, among whom may be mentioned Leo Daft, who, in 1883, operated the Saratoga and Mount McGregor Railroad by electricity. Edward M.

Bentley and Walter H. Knight also deserve mention for their pioneer work, which tended mainly in the direction of supplying a practical system for operating railways by means of the conduit system; and finally C. J. Van Depoele, to whom the progress which the electric railway has made in this country is largely indebted.

With this brief review of the efforts which have led up to the elec-

Fig. 1.—Edison's Menlo Park Electric Locomotive, 1880.

tric railway of to-day, I shall pass to the consideration of the subject as it presents itself in its latest aspect.

Broadly speaking, the electric car is a self-propelling vehicle, in which the propelling force is furnished by a motor actuated by an electric current. For the purposes of convenience, electric railways may be divided into three classes, depending upon the manner in which the current is supplied to the electric motor upon the car. These are:

1. The "*overhead system*," as it is called, in which the current is led from the generating machine at the station to the car through a wire placed above the ground.

5

2. The "*underground system*," or that in which the supply conductors are placed below the ground.

3. The "*storage-battery system*," in which the current is furnished by storage-batteries carried on the car, which have been previously charged with the required current.

Though differing in name, these various systems are alike in principle, and, indeed, have much in common; but this artificial distinction may be conducive to a better understanding of the subject.

As other chapters * in this book give the reader a sufficiently good idea of the theory and action of the electric motor and the dynamo,

Fig. 2.—Plan showing Principles of Operating the Overhead System of Electric Railways.

they need not be here described, and a view of the plan upon which the first of the systems of modern electric railways above mentioned is operated can be at once presented. The sketch, Fig. 2, shows in outline the principal elements of this system. These consist, broadly speaking, of the generating station, the line, the car, the motor, and the return-circuit. At the generating station there are an engine and boiler which furnish power to drive the dynamo *D*. The current generated by this machine is conducted by a wire to the line *L*, which is strung on posts and runs parallel with the track. The car, in order to obtain the current, makes continual contact with the line *L* by means of a trolley, the current passing down by wires to the motor *M*, connected with the axles of the car. After passing through the motor, the current passes into the wheels

* See "Electricity in the Service of Man," and "Electricity in Lighting."

of the car, and **thence** into the **track ; the** latter, it will be seen, is connected to the other pole of the **dynamo** *D,* and a complete circuit is thus formed. It will be noted that in addition to the track connection as a return for the current, the earth is also called into play, acting as a conductor in the same manner as it is employed in telegraphy, and with the same advantages. This is effected by connecting the track at intervals with large plates shown at *E,* buried in the wet ground, and the integrity of the circuit is additionally enforced by connecting the rails electrically by means of copper wire, indicated at *J,* as the ordinary fishplates joining the rails cannot be relied upon to give a continuous electric circuit such as required.

Some of the more important details, upon the success of which the operation of the electric railroad largely depends, should be next considered.

As recently remarked, with much truth, by a writer in referring to the electric street-car: "The truck is the car;" hence, as this element is common to the three systems above mentioned, it seems first in order to claim attention. The truck being the support of the car-body in which the passengers are carried, is necessarily limited to certain dimensions, and the problem of concentrating motors of sufficient power to propel the car, into the limited space available, afforded a good field for inventive genius. Again, the manner in which the power generated by the motor was to be transmitted to the wheels and axles, though apparently simple, was found to be by no means easy of solution ; and even at the present time differences of opinion exist on this point. Economy in weight as well as in power requires that motors shall be run at high speed, and, as the car-wheel, as a rule, runs at comparatively low speed, it is evident that some method of reducing the speed of the motor to that of the car-wheel must be employed. Among the various methods which have been proposed and tried are friction gearing, connection by means of belts, the sprocket and chain, the worm and wheel, the direct crank action, and finally the gear and pinion. Of all these, the last

may be said to be practically the only one which has thus far come into
any extensive use, at least so far as this country is concerned ; and, as
the number of our railways in operation far exceeds that of all the rest
of the world put together, it is safe, for the present at least, to designate
this method as the typical one in use to-day.

In order that the reader may therefore clearly understand the con-

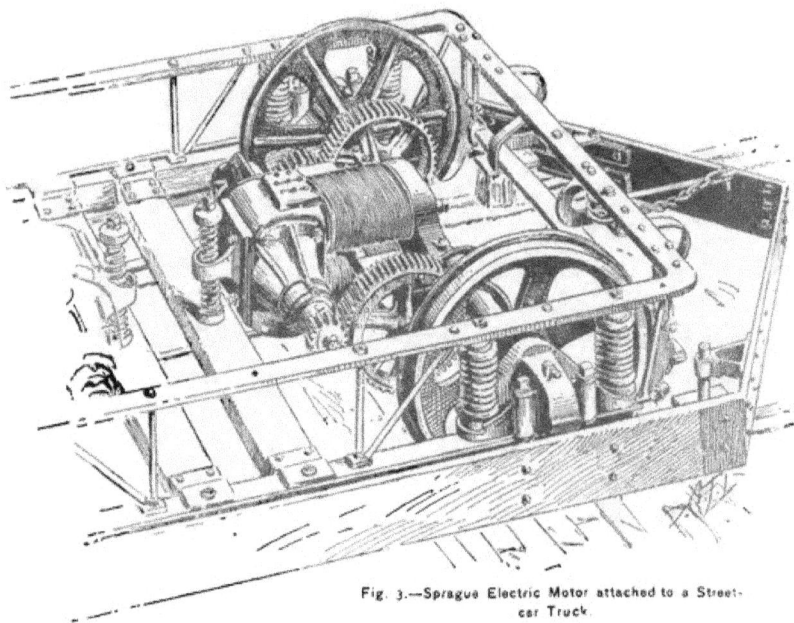

Fig. 3.—Sprague Electric Motor attached to a Street-
car Truck.

struction of the ordinary electric railway truck, a view is shown in Fig.
3 of the form designed by Frank J. Sprague, one of the most successful
of the new school of electrical engineers. As the space between the bottom of the car and the ground is necessarily confined, it has been found
expedient in practice to divide the motive power into two units by the
application of two motors, one to each axle, as it is evident that one
motor sufficiently powerful to do the work would, as a rule, be very

difficult to place under the car without interfering with its present construction. The manner in which the power of the motor is transmitted to the wheels is very clearly shown. The only moving part, the armature, has at one end of its shaft a small gear-wheel which meshes with a pinion placed upon a counter-shaft which passes through the legs of the magnet; and the other end carries a similar pinion, gearing with a toothed wheel connected to the axle of the car. Hence the armature of the motor, which runs at high speed, transmits its power to the axle at a lower speed by means of this gearing. The successful operation of this gearing, however, requires that all these wheels shall remain in a constant fixed relation to each other, and in order to accomplish this the very ingenious expedient has been applied of centring the motor itself upon the axle of the car; thus, no matter how much the vehicle may be jarred during its passage over the track, these wheels will always bear the same relation to each other and to the axle upon which they are mounted—a most essential point for their proper operation.

Provision must also be made for the easy starting of the car, and to prevent disagreeable shocks from the sudden starting of the motor when the current is switched on. This is accomplished by suspending the free end of the motor between a pair of springs, which are shown supported by cross-bars stretching from side to side of the truck. Thus the motor is given free vertical play for a short distance, and the shocks which would be caused by a rigid arrangement are taken up by the springs and the car started with a gradual movement. It may be said that the advent of the electric railroad has entailed an entire remodelling of the street-car truck formerly employed, and has indeed constituted an almost distinct, new field of invention.

It is upon a truck of the nature above described that the car-body is mounted, and the result of the construction adopted is that the working mechanism is entirely removed from view.

A small but very important detail, which has added much to the successful operation of the motor and the car, consists in the substitution of a carbon brush bearing against the commutator, in place of the copper brush, which had until very recently been employed. Small as this detail may appear, it is almost safe to say that it constitutes one of the most distinct advances in the electric railway motor that has been effected since its practical application.

Another very interesting method that has been proposed for transmitting the power of the revolving armature to the axles and wheels consists in mounting the armature directly upon the axle of the car, so that no intermittent gearing whatever is required, the armature shaft and the axle being identical. A late idea in this department is embodied in the design of Mr. William Baxter, Jr., of Baltimore, who proposes to enclose the motor entirely within the car-wheel, and thus to relieve the axles of all strain due to the weight of the motors.

The most recent development in this branch indicates the employment of slow-speed multipolar motors, by the use of which one pair of intermediate gears can be avoided.

The consideration of the various methods by which the current is led from the generating source to the motor on the car, by means of the overhead wire, can now be entered upon. This, evidently, is most important, as upon the effectiveness and integrity of the " line " depends the successful operation of the road, just as in telegraphy the line wire requires to be maintained perfect in order to effect communication.

Looking back to the early electric railways operated by Siemens at Berlin, it is found that the same arrangement, long practised in telegraphy (which is depicted in Fig. 2), was there adopted; but the conductor, instead of being overhead, consisted of a central rail placed between the other two, but insulated from the ground. The current from the dynamo first passed through this central rail, then into the motor through the wheels, and then into the two outer rails and the ground, which carried it back to the other pole of the generating dynamo.

This construction was also adopted in his early work by Leo Daft, in this country; but it is evident that, except in special situations, it is not suitable on account of the danger of shock which it involves to persons and animals crossing the tracks, by coming in contact with the conductor. The two rails themselves have also been employed exclusively as conductors, the one rail being the positive side of the system, and the other the negative.

The overhead line of to-day, in connection with electric railways, is going through the process of evolution similar to that of the other elements of the system. The first attempts in this direction consisted in fixing upon posts a tube having a slot running along its entire length, and facing downward. Within this tube there was placed a slider, which was connected to the motor on the car, and which served to maintain a continuous contact between the moving car and overhead conductor. The operation upon this method, though still continued in one or two instances abroad, was soon abandoned, however, and its place taken by the plain cylindrical wire upon which a trolley-wheel was maintained, which moved in connection with the car, and served to make the necessary contact between the motor and the overhead conductor. This trolley had therefore necessarily to be supported by the wire, and consequently demanded a wire of suitable strength to stand the strain of the travelling wheels. Hence, to avoid this difficulty the very ingenious idea was adopted of supporting the contact-wheel at the end of an arm resting on the top of the car, and pressing it in contact with the lower side of the wire; as a result of this it is evident that the wheel, instead of being a load upon the wire, actually serves to support the wire in its course; and, consequently, a much lighter construction can be adopted in this case than in that previously mentioned.

The manner in which the conductor carrying the current is maintained in position overhead is subject, naturally, to the conditions both of the traffic and the nature of the road through which the tracks pass. Therefore there are various types of overhead constructions. In ordinary cases, in cities where two tracks are placed side by side in a street, there are two general modes of suspending the overhead wire. A very admirable example of the manner in which this can be accomplished, without obstructing the street or in any way marring its beauty, is that which is illustrated in the engraving, Fig. 4, which represents the Thomson-Houston electric railway, operating in Washington. Here ornamental iron poles are placed at suitable intervals, and carry cross-arms, from the ends of which the wire is suspended by means of an insulator.

Fig. 4.—Overhead Wires for Double-track Road, suspended on a Single Line of Ornamental Poles.—
Thomson-Houston Railway, Washington, D.C.

This simple construction permits also the illumination of the street—for it may be noted that every second pole is surmounted by a cluster of incandescent lamps which light up the roadway both for the cars and for the traffic which may be passing on the streets. These lights may be run from the same current which supplies the motors on the cars;

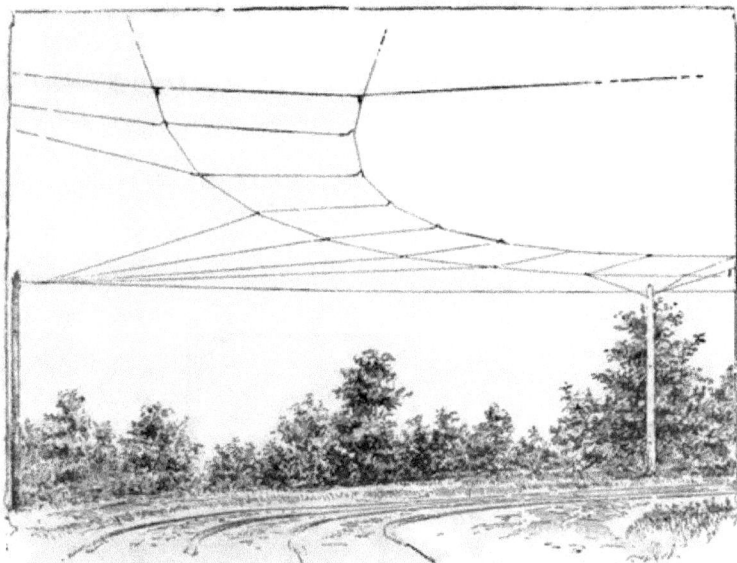

Fig. 5.—System of Overhead Wires, suspended from Poles on opposite sides of the Street. The above a curve on the Sprague Railway at Wilkesbarre, Pa.

but where this is not considered desirable, a separate conductor can be strung for that purpose; in either case the posts themselves afford a ready means for the suspension of the lamp.

Where the streets are not wide enough to permit of the adoption of a system of poles running along the centre, another method is frequently adopted, which consists in placing the posts at the curb line on either side of the street, and suspending the conductors by means of wires stretched from opposite poles across the street. This method of construction is shown in the engraving, Fig. 5, which represents the operation of the Sprague electric railway at Wilkesbarre, Pa.

Electric railways in many instances connect cities with their suburbs, with tracks frequently running for considerable distances. The method of overhead construction in such cases consists in using a line of poles having single arms extending from one side, a general type of

which is well illustrated in the engraving, Fig. 6, which shows a section
of the Thomson-Houston electric railway at Rochester, N. Y.

In the outline sketch, Fig. 2, the main conductor is represented by
a single wire. It is evident, however, that any break in the overhead

Fig. 6.—Poles with Single Arms for Suburban Roads.—The Ontario Beach Railway, Rochester, N. Y.

circuit, as there shown, would cause an interruption to the traffic.
Hence at the present time there are, in fact, two systems of conductors
employed; one of these, called the main conductor, is run out from the
dynamo generating station to various parts of the road, and connects
with the working conductor, as it is called, to which the trolley-wheel
makes contact. The working conductor being thus fed into at a dozen

places, a break in any one part of the circuit will not cause any interruption of the current, so that in reliability of operation the electric railway is far superior, probably, to any other method now in existence, and indeed much preferable to the cable railway, in which the operation of the road depends entirely upon the integrity of the cable, and any stoppage of which means a total interruption of traffic.

As simple a matter as it may seem, the successful operation of the "under-contact" trolley required an enormous amount of experimentation before the proper type of contact was obtained. The one in general use to-day consists merely of a grooved wheel, which is fixed at the end of the trolley-arm. As there is always more or less sag to the wire, some method must be provided for keeping the wheel in constant contact, which evidently could not be effected if the wheel were rigidly attached to the car-body. To effect this, therefore, the arm upon which the contact-wheel is mounted is pivoted flexibly to the top of the car, a series of springs serving constantly to push the arm upward. It is at the same time sufficiently yielding to allow it to overcome any inequalities in the level of the wire or of the road. The arrangement is such that the arm has a free motion from a vertical position to a perfectly horizontal one, so that electric cars may pass under bridges, for instance, reaching to within six inches of the top of the car.

The overhead system so far described consists practically of but a single overhead wire, with a ground return for the current; but there are still some who prefer to use a continuous metallic overhead circuit. This naturally entails the running of two wires instead of one; one wire serving as a feeding wire and the other as the return wire. The principle of operation is evidently the same in both cases, and a very interesting example of this case of overhead construction is that afforded by the Daft electric railway operated in Cincinnati, a view of which is shown in Fig. 7.

It may be remarked that, although the large majority of the roads in operation to-day make contact with the under side of the wire by means of a wheel, there are still some who adhere to the older practice

of maintaining a sliding contact with the conductor; among them being
Sidney H. Short, who prefers a sliding contact at the end of the arm
which is pressed up against the under side of the wire, and continually
rubs against it in its passage.

The adoption of the overhead system has been so general that but
comparatively little has yet been done in the way of a practical appli-

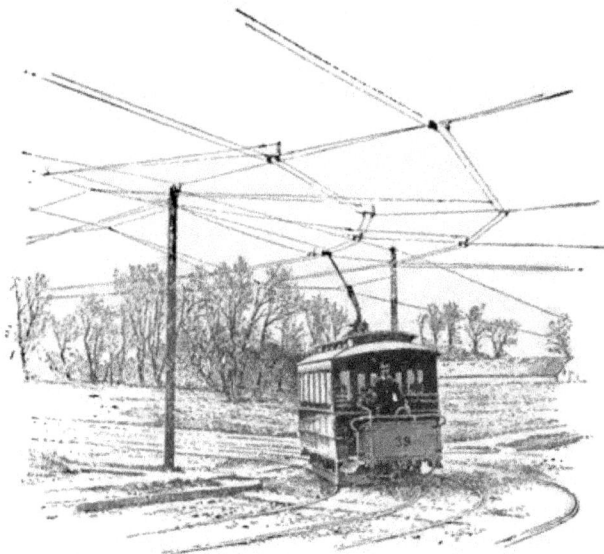

Fig. 7.—The Double-wire, Continuous Metallic Circuit System.—Daft Railway, Cincinnati, O.

cation of running conductors underground. It is evident that by maintaining the system above the ground, it can be closely watched and
readily inspected at all times, and the slightest fault which may be developed can be hunted up and remedied in the shortest possible time.
Again—and perhaps this may be deemed the most important factor
which has led to the preference of the overhead system to the underground—there is the small cost at which it can be erected and maintained.

But it was early evident that the demand of the public in crowded cities would in time force the adoption of some underground system, and various plans have been suggested with this end in view. Evidently the principle remains the same as that employed in the overhead system, but many are the difficulties which present themselves when the conductors are placed below the surface. The problem involves, in the first place, a construction which will effectually resist the action of all forces tending to disturb the relative position of the wires underground; and where the traffic on the streets is very heavy this involves a very strong construction of the conduit. Again, it is absolutely necessary that the conductors shall remain thoroughly insulated from each other and from the ground under all conditions of weather. The frequent heavy rains and snows occurring in this country, therefore, necessitate the adoption of a construction which shall permit of a thorough insulation of the conductors and a drainage of the entire system, not only to prevent an entire stoppage of the operation of the road by flooding, but also to avoid a continuous loss of current from conductor to conductor by leakage. To such general conditions are added others of minor importance. To meet all these, therefore, has been the subject of not a little study. Only a comparatively brief reference can be made to one of these types, the design of Messrs. Bentley and Knight. It is well illustrated in the engraving [Fig. 8]. A number of constructors have arranged the conduit to run along the centre of the track, but the objections to this method of operation have been overcome by placing the conduit at one side of the track. As shown in the engraving the two conductors are supported upon porcelain insulators fixed to the sides of the conduit. Placed directly above them are the two slot-rails through which a plough attached to a cross-beam on the car-truck enters. The lower end of the plough carries two contacts mounted upon springs, so that they are kept in continual contact with the conductors. The conduit is constructed of heavy cast-iron, horseshoe shaped ribs, which are laid in the ground and connected continuously by an iron shell fixed to the flanges. For the proper and easy examination of the conduit, hand-holes are provided at short intervals, one of which is shown in section in the engraving.

Although a limited number of electric railways operating with the conductors placed in conduits are in successful operation to-day the difficulties encountered in their operation have led inventors to seek other means of communication between the conductors and the motor than is afforded by means of a plough; moreover, a slot running along the surface of the street is also looked upon by some as an objection, the removal of which would be desirable. To avoid both these a number of inventors have hit upon the idea which consists in laying the

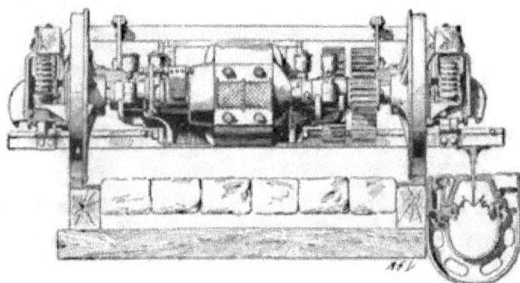

Fig. 8.—The Bentley-Knight Underground Conduit System, showing Cross-section of Track, Conduit, and Truck.

conductors underground, and, at short intervals, providing devices which shall close the conductor circuit through the car at whatever place the car happens to be. In one of these systems, that designed by Messrs. Pollak and Binswanger, a magnet carried at the bottom of the car acts upon a switch placed, every twenty or thirty feet, below the surface of the street, which switch closes the circuit and sends a current through the motor on the car from the main conductors. A system of a similar nature has also been designed by Mr. McElroy, of Pittsburg. Quite recently a system on this plan has been put in operation near London by Mr. Lineff.

As remarked recently by a well-known electrician, the underground electric-railway problem does not of itself present any inherent difficulties, but an essential element in its success is proper engineering, such as has been proved necessary as the result of past experience in cable traction. And, according to the same authority, the laying down of a cable conduit ought to be hailed with delight by electricians, as, sooner or later, it will most probably serve as a receptacle for electric-railway conductors.

There are two different ways in which electric cars may be operated, considering their electrical relation to the conductor. Electric lamps, as is explained in the chapter on "Electricity in Lighting," may be connected so that the current passes through each lamp in succession. This is the system upon which the large arc lamps for street illumination are connected, and is called the "series" system. Another method, however, which is that almost universally employed in connection with the incandescent lamp, is the connection of the lamps across the circuit, the lamps being, as it were, placed parallel to one another across the outgoing and returning wires, like the rungs on a ladder, and each lamp obtaining its current independently of the other. This is called the "multiple arc" or "parallel" system. The latter method is the one upon which the large majority of electric railways running to-day are operated. It requires that the electric pressure at the terminals of the dynamo, and hence upon the line, shall remain constant, while the current passing over the line varies, of course, with the number of cars which are being operated at the time; ten cars, for example, taking ten times as much current as one car.

But the series system of operating cars still has its adherents, among them Sidney H. Short, of Cleveland, O. In his system the current is maintained at the same strength throughout, and passes from one car to the other undiminished in strength. It involves, however, a change in the electric pressure of the line, so that with ten cars the pressure would be ten times as great as that required with only one car in operation. Thus, although no actual power is saved, since in one case the pressure, and in the other the strength of the current, is varied proportionately, its adherents claim for it certain other advantages in operation, among others, a saving in the cost of conductors.

The storage-battery system is frequently called the ideal system of street-car propulsion. It is true it is the most pleasant to contemplate both from the standpoint of the public and the street-car manager. The objections which are held against the erection of wires and poles in streets, or the placing of conduits which necessitate slots in the road- way, would evidently be entirely overcome by a system which should leave each car independent of every outward source of power. This great desideratum is undoubtedly best embodied in a car equipped with its complement of storage-cells previously charged, the car moving over the road as a single unit independent of all other conditions. These manifest advantages were early recognized, and hence it was not long after the practical storage-battery was invented by Faure, that attempts were made to apply it to street-car propulsion. The first of these cars was put in operation in Paris, in 1882, and was followed by experimental operations in various other places. In 1885 a competition at the Antwerp International Exhibition, arranged between an electric car, steam locomotives of various kinds, and a compressed-air engine, resulted in the complete victory of the first. Progress has, however, been steadily going on, and though but few such roads are in operation as compared with their more vigorous competitor, the overhead system, the belief is entertained by many that, with improvements that will undoubtedly be made in the storage-battery, this system will occupy a very prominent position in the future of electric traction. The reason for this will be apparent when we consider the very simple elements of which it is composed. The motive equipment of the car does not differ essentially from that already described in connection with the overhead system; but to this is added a set of storage-batteries which hold a suf- ficient charge to propel the car a given number of trips. The illustra- tion [Fig. 9] shows such a car as operated at present by the Julien Electric Traction Company, in New York. The batteries are placed under the seats, and occupy no space otherwise useful. This system requires, of course, like those above described, a station in which a sufficient current is generated for charging the cells. Here the cells are charged in regular rotation; the car, after its run, enters the car-house, discharges its exhausted cells, and is furnished with a new set, which have in the mean time been charged. This operation requires but a minute or two. The arrangement can be so made that the work of the engines at the station in charging the cells is practically continuous

Fig. 9 —The Storage-battery System.—Car of the Julien Electric Traction Company, as run on the Fourth Avenue Road, New York.

during twenty-four hours if necessary, which conduces to a well-known economy in operation.

Such, in general, are the main features of the systems of electric railways which have thus far been developed to any considerable extent. The rapid extension of the electric street-car system which has taken place (especially in this country), naturally leads to the question of the cause thereof. To have gained such preëminence it must be able to do not only what other systems can do, but, still more, it must be able to do it at a decreased cost. Again, removal of thousands of horses from the streets of a city, involving, as it does, the doing away with the noise and dirt, is another distinct gain to its residents. But if

6

one goes still farther, and contemplates the difference between a stable
housing thousands of horses, and an electric-car station of sufficient
size to operate a road with the same efficiency, one is at once struck
with the advantages on the side of the electric system, which, indeed,
are incontrovertible. Instead of a large, ill-smelling building whose
odors are wafted for many blocks (making the tenancy of houses within
half a mile almost unbearable, and involving a large depreciation of
property in the neighborhood), there is a neat, substantial building
equipped with a steam plant and dynamos, and occupying hardly one-
tenth the space required for an equivalent number of horses. There
fore, not only is there effected a removal of the nuisances attached to a
stable, but a large saving in the cost of real estate, and the far greater
amount involved in the known depreciation of the surrounding property. Besides this, the stables are of necessity required to be in close
proximity to the track, whereas the electric power station, which furnishes current to the car, may be situated a mile from the track in some
suitable place, as, for instance, beside a river, where, with condensing
engines, power may be generated at a minimum of cost.

Again, looking at the electric street-car from the standpoint of the
engineer, it becomes evident that it is an undisputed rival of all other
systems of mechanical propulsion. For example, it requires no device
for the suppression of dirt, dust, and smoke in the streets, the necessary accompaniment of all steam locomotion. But most important of
all is the consideration that the electric motor has, in fact, but a single
moving part, the armature, the motion of which, unlike that of the
steam and compressed-air engine, instead of being reciprocating, is
rotary, and hence avoids the disagreeable jolting which attends the
riding in cars which are of necessity frequently required to start and
stop. As a consequence of there being but a single moving part, the
cost and care required to keep the electric motors in running order is
but a minimum, and the art of building them has to-day advanced to
such a point that, with intelligent supervision, the life of the machine
is equal to that of any similar mechanism.   -

It is fair to assume that but few roads exist which are so favorably situated that they encounter no grades in their course; and when the proposition to employ electricity as a traction agent was first projected, the difficulty as to the ascent of grades was held out as one of the drawbacks to the application of the system. But it required but a short period of actual experience to demonstrate that in just such situations the electric car was superior in every respect to the horse, and indeed to the steam locomotive. Grades exceeding ten per cent. are being overcome on roads now in operation, and others of lesser degree are now considered as of easy accomplishment with the electric car. In order to be able to cope with such grades it is, of course, necessary that the motor attached to the cars have ample power, and it has therefore become the custom to equip the trucks with two motors ranging from 10 to 15 H. P. each, thus giving the car an available traction power of from 20 to 30 H. P. Considering the fact that the ordinary horse-car has, as a rule, but two horses, this might to some appear an excessive amount of power equipment; but the fact must not be lost sight of, that while, ordinarily, two horses exert their normal, average strength in keeping the car in motion when once brought to its proper speed— the effort which they exert in bringing a car from a dead standstill to its proper speed often actually exceeds ten horse-power. Hence it is that the frequent heavy exertion required of horses in the street-car traffic results in their rapid wearing out and final disability for active service after three or four years' work. Therefore it is necessary that the electric car should be provided with the power corresponding at least to that which the horse exercises when required; but it is evident that, once started, the motor need only deliver a small part of its capacity, sufficient to keep the car in motion. But since electric cars are put upon roads having grades which have not been attempted with animals, additional power is frequently required, and hence it is that as high as 30 to 40 H. P. are sometimes concentrated on one car which, under normal conditions, hardly requires more than three or four for its propulsion. This increase of power has also been necessitated by

the practice which has sprung up of coupling one, and sometimes two
or three tow-cars, with a motor car, so that in reality the motors of one
car are required to do the work of two or three.

In this connection attention should be called to a phenomenon
which may now be considered to be an established fact, in virtue of
which electric cars are aided in ascending heavy grades. This phenomenon, which was probably first observed by Leo Daft, at his works in
Greenville, N. J., in 1882, is that, when the current passes from the car-
wheel to the track it causes an increased friction or resistance to sliding
between them, the result of which is that slipping is to a large degree
prevented, and heavier grades can be attempted: and, on the other
hand, heavier loads taken up than would be practicable with a system
in which the current did not pass between the wheel and the rail. The
explanation of this phenomenon, though not completely established,
seems to lie in the direction of a slight welding action which takes place
between the wheel and the rail, caused by the heat generated by the
current.

In respect to the regulation and operation of electric cars, it may be
remarked that there is no system which is more elastic. The driver at
the front of the car has under his control the switch, so that by a simple
movement of a handle he may regulate at will the speed of the car
from a standstill to full speed, as well as its direction of motion. Up
to the present time the hand-brakes, as a rule, have been retained; but
it is evident that with a motor under the control of a driver which can
be instantly reversed, a powerful addition to the ordinary hand-brake
is placed in the hands of the driver, and this has been often turned to
good advantage to prevent accidents. In support of this it may be
cited that since the inauguration of the electric railway in Cleveland,
O., the number of accidents has been far less than for the corresponding period during which the road was operated by horses, notwithstanding the fact that the electric cars are run at a higher speed.

The operation of street railways by electricity, although even now completely demonstrated to be more economical than by either horses or cables, is yet too recent to afford the more reliable figures which can only be obtained after extended use; but from an investigation made on a number of roads by O. T. Crosby, some very interesting data are developed. The results of Mr. Crosby's investigation show that the average cost of motive-power for the roads in Washington, Richmond, Cleveland, and Scranton, was about 5.09 cents per car mile, and the relations of the various items which go to make up this total cost are exceedingly interesting. Thus it is shown that the interest on the investment constitutes about one-fourth or one-fifth of the whole; that is to say, about one cent per car mile; coal, as a rule, about twelve per cent.; attendance, about forty per cent.; and the machinery and line. without interest, the remaining twenty per cent. But with all these manifest advantages of the electric railway, the best proof of its superiority is to be found in the experience of those who are using it; and if the unsolicited praise from that quarter is to be relied upon, then certainly the electric railway is an unqualified success.

At the eighth annual meeting of the American Street Railway Association, held September, 1889, at Minneapolis, the committee which had been appointed for the purpose of investigating and reporting upon electric railways, submitted a report which should finally set at rest the doubts of those who still believe the electric railway to be in the experimental stage. This committee reported in fact that, "if it is desired to make a change from horse-power, electricity will fill the bill to perfection, no matter how long or short the road, or how many passengers are carried. In the investigation of the subject the most satisfactory results have been shown; it not only increases the traffic over the road, but reduces expense, and actually enables us to operate a line, which heretofore entailed a loss, at a profit." After discussing the various systems, the committee gives an estimate relative to the cost of equipping a railway on three systems, namely, on the cable system, the overhead wire, and the storage-battery system, which is as follows:

A comparative statement of the cost of construction of a ten-mile road complete, with fifteen cars, would stand probably as follows :

Cable System :

Cost of cable construction.............	$700,000
Cost of power plant..	125,000
Cost of cars..	15,000
	$840,000

Electrical Overhead-wire System :

Cost of road-bed... .	$70,000
Cost of wiring...	30,000
Cost of cars..	60,000
Cost of power plant..	30,000
	$190,000

Storage-battery System :

Cost of road-bed...	$70,000
Cost of cars...	75,000
Cost of power plant......	30,000
	$175,000

In the above cases of electrical construction, the motor-car would be capable of pulling one or two tow-cars, if necessary. These figures your committee have no doubt will be found to be calculated within a reasonable limit of cost.

Here, then, is at once a most potent argument for the adoption of the electric railway over the cable system, for (while answering all the demands which can be made upon a car) its cost of installation is nearly five to one in favor of electricity. To this must be added the fact that in the case of the cable, under favorable conditions, only eighteen per cent. of the power of the engine is actually employed in the propulsion of the cars, the remainder being consumed in the mere haulage of the dead cable; while in the electric system at least fifty per cent. of the engine power is available for traction purposes. The cost of power, or coal required, is thus approximately three to one in favor of electricity.

As remarked in that report, the installation of an electric railway in place of horses is uniformly accompanied by a large increase in receipts, as well as a decrease in expense. Both of these items working together have resulted in a most remarkable showing of earnings for such roads. Only a few instances need be given to demonstrate this: The electric railway at **Davenport, Ia., started** on September 1, 1888, with five four-teen-foot cars. **The road included a grade of seven and a half per cent.** for sixteen hundred feet, and the following table gives a comparison of the earnings for four consecutive months, operating with horses and with electricity:

	1887.		1888.		Net increase per cent.
	With horses.		With electricity.		
	Gross.	Net.	Gross.	Net.	
September	$1,347 49	$474 79	$1,907 15	$997 15	110
October	1,232 47	302 47	1,903 94	1,121 94	270
November	1,131 49	231 49	1,886 06	986 06	320
December	1,283 14	353 14	2,022 98	1,123 48	220
	$1,255 40	$340 47 Aver.	$1,952 53	$1,056 91 Aver.	210 Aver.

As here shown, there was an **average** net increase of **two hundred and ten** per cent. in the receipts. Other places have shown still more remarkable results, but the reticence of the managers of these roads naturally prevents the publication of what might otherwise almost be considered apocryphal earnings. One case may be mentioned in which, for thirty-one days, during the month of **July, 1889, the receipts** amounted to $10.605, and the operating **expenses $3,735, showing a net** gain of $6,870; and another in which, for the month of August, 1889, the operating receipts were $4,317.46, while the total expense amounted to $871.04, giving a net profit of **$3,446.42.**

The popularity which the electric cars have obtained in cities where they have been employed is well known, and easily accounts for the remarkable showing made in the earnings of the road. The service, instead of being slow and uncertain, as under the régime of the horse, is now swift and sure, and delays are practically unknown. For a time doubts were expressed of the ability of the electric cars to cope with the conditions imposed by our harsh Northern winters, but the experience of the last two years has shown that such fears were unfounded, and the most severe storms which passed over this country last winter caused not the slightest delay in the operation of electric railways. A good example was given of this immunity from delay by many of the Western roads, among them those of Omaha, Council Bluffs, Cleveland, Davenport, and St. Joseph, where the electric cars maintained schedule time, whereas the horse-cars were running at irregular intervals with double teams. It is evident that with a sweeper provided with powerful motors for removing the snow from the tracks, and kept constantly running over the line, there is nothing to prevent its being kept clear at all times. Even without the sweepers, the cars themselves have sufficient power to force the snow aside and maintain the track clear, as has often been demonstrated.

Our own country has made far greater progress in the application of electricity to railways than all the rest of the world included, and it is therefore not uninteresting to glance briefly at the rapid increase which the system has undergone. The first trustworthy statistics on the subject were given in a paper read by T. C. Martin before the American Institute of Electrical Engineers, in May, 1887, in which he showed that there were in operation at that time in the United States thirteen electric railways, carrying about three million five hundred thousand passengers annually. The latest and most trustworthy statistics relating to the same subject show that there are in operation in this country, and in course of construction at the present time, no fewer than 260 electric railways, operating over 3,000 cars, with 1,753 miles of track. The number of passengers carried it would be difficult to estimate, but it must be considerably more than 1,200,000 daily.

Among the larger cities in which electric railways have been put in operation, the foremost is Boston. W. H. Whitney, the president of the West End Railway, of Boston, after thorough investigation and trial of the electric railway, was finally so well convinced of its superiority over all other methods of street-car propulsion that he recommended its general adoption on the street railways of Boston ; and while more than two hundred cars are in operation there at present, preparations are going on which will culminate in the operation of nearly one thousand electric cars in Boston alone. Among the other cities having electric railways is Cincinnati, with forty cars, and preparations for a large increase. Cleveland, O., has now several lines operated by electricity, as well as Harrisburg, Pa.; Kansas City. Mo.; Hartford, Conn.; New York City; Omaha, Neb.; Pittsburg, Pa.; Salt Lake City, Utah; San José, Cal.; Scranton, Pa.; St. Louis, Mo.; Tacoma, Wash.; Washington, D. C.; Wilkesbarre, Pa.; Wilmington, Del., and a long list of others.

Wherever the electric railway has been introduced a reduction in

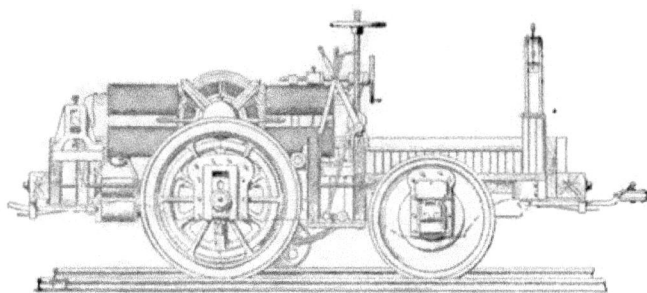

Fig. 10. Daft Electric Locomotive for Traffic on Elevated Railroads.

the schedule time, or, in other words, an increase of speed, has followed; and where the lines connect the suburbs of cities, not infrequently a speed of from twelve to eighteen miles per hour is attained by electric cars, thus affording to residents in suburbs the speed facilities of a steam railway.

For intra-urban rapid transit, evidently, electricity is superior in every respect to steam traction, and hence it was but natural that several electricians should have essayed the solution of the problem of affording the residents of New York a deliverance from the present overcrowded condition of the elevated railway cars. Among the electricians who have submitted plans for this may be mentioned Leo

Daft, who was the first to place an electric locomotive on the elevated
railroad, and who has recently shown, as the result of his experiments,
that he is able to increase the traffic of the road with a reduction in
cost of operating expenses. The locomotive employed by Mr. Daft in

Fig. 11.—Stephen D. Field's Motor.—Experimental Trials on the Thirty-fourth Street Branch of the
Elevated Railway, New York City.

his latest experiments, called the "Ben Franklin," is shown in elevation
in Fig. 10.

Frank J. Sprague has also attacked the problem, his plan embodying the idea that the locomotive car shall also be a passenger car, only
about one half of its total length of fifty feet being occupied by the
motive-power equipment. In this way the weight of the locomotive is
widely distributed over the road-bed, a necessity with the present form
of elevated railway structure.

Stephen D. Field has also turned his attention to this problem, and like Leo Daft, favors the employment of an electric locomotive independent of the rest of the train. His motor, as run on the Thirty-fourth Street branch of the elevated railway in New York City, is illustrated in Fig. 11, and embodied a modification in the gearing of the motor from those heretofore employed. It will be seen that instead of employing intermediate toothed gear, or a similar device, Mr. Field connects directly to the armature shaft a crank which, through the medium of a connecting bar, transmits its motion directly to the wheels of the locomotive.

Though the experiments undertaken on the elevated railways have not yet led to the adoption of that system, it is only a question of time when it will become a necessity, and, indeed, the only way out of a constantly increasing difficulty. The elevated railroad presents ideal conditions for the application of such a system, and the cause of the delay which has thus far taken place must be looked for rather in a conservative management than in any lack of appreciation of the proposed system.

The advantages of the electric railway on the surface of the earth have been pointed out, but by those who have ever witnessed the operations of a railway within mines. the introduction of the electric locomotive will be admitted to be one of the most marked advances which have been made in that industry during recent years. Indeed, one of the first electric railways ever operated was a mine tramway. Removing at once the slow and obstinate mule on the one hand, and the dust, smoke, and noise and poisonous gases of the steam locomotive on the other hand, the electric locomotive does its work with "neatness and despatch," requiring but a fraction of the attendance necessary in the other methods, and promoting the comfort of the miner in the highest degree. The ingenuity of the electrician has easily adapted the electric motor to these purposes. A mine locomotive employed at Scranton, Pa., by the Hillside Coal Company, designed by C. J. Van

Depoele, has shown itself fully able to handle several hundred cars per day, and has entirely displaced the mules formerly employed in the mines. Several other mining railways are running, or in course of equipment, in this country and several are in operation in Europe. This mining branch of electrical development, though hardly touched at the present time, is certainly destined to equal, if it does not exceed, in extent the wonderful growth of the surface railroad.

Inventive genius early in the art looked to a further extension of electric traction, and as early as 1882 Professor Fleeming Jenkin suggested the idea of an electric transportation system in which the motor

Fig. 12.—The Glynde, England, Telpherage Line. on the System of the Late Fleeming Jenkin.

or car should ride upon a suspended cable, which should at the same time constitute both the track and the electrical conductor. This system, which was named by him "telpherage," has actually gone into operation at Glynde, in England, where it is employed in delivering clay from the mines for a distance of several miles. This system is illustrated in Fig. 12. The great cheapness of this system of construction, together with its flexibility, seems to promise for it a bright future. The train is under complete control of the attendant at the station. As a feeder to the main railway lines of traffic it possesses unquestionable advantages, and for the transportation of ore, coal, and minerals generally, as well as corn and other agricultural products, it would seem to have many advantages.

Fig. 13.—The Weems System for a High-speed Electric Railway. More than one hundred and twenty miles an hour actually accomplished on an experimental Track.

These descriptions have thus far been confined to what has actually been accomplished; but it is not out of place to cast a glance into the future, in order to discern in what direction electricians are working in the domain of electric railways. One of their main objective points is to attain higher speed than is now reached with the fastest express train, and enough has already been demonstrated to show that this is by no means impossible. There was for some time in operation at Laurel, Md., a system of electric railway, originally designed by David G. Weems. When it was recently inspected by the writer, with his watch in hand, he noted a speed of the electric locomotive of nearly one hundred and twenty miles an hour. The electric car there employed is illustrated in Fig. 13. The electric motors are constructed with a revolving armature which is mounted directly on the axle, so that no intermediate gearing whatever is employed. The curiously pointed ends of the car, which might by some be considered fantastical, have their *raison d'être* in the fact that, at the high speeds at which this car is run, the resistance of the air is by far the greater retarding influence; much greater, in fact, than the resistance due to the axle and rolling friction, which at lower speeds is predominant. The electric current is taken from a conductor fixed above the car, to which a brush connected with the motor makes contact. There is certainly nothing in the new system which could prejudice its feasibility under suitable conditions.

There is also another system of rapid transportation which has been suggested, and has been put into experimental operation, known as the "Port-electric" system. In this system, invented by John T. Williams, a well-known principle is applied, namely, that of the sucking in of an iron core by the action of a current circulating in a coil around it. Mr. Williams makes his car or carrier play the rôle of an iron core, which is propelled by the successive action of coils of wire placed at suitable intervals along the track.

With the advantages of the electric railway so clearly pointed out, and so unquestionably demonstrated in actual practice, it would not be unsafe to hazard the opinion that, in ten years at the farthest, there will not be a *single* horse-railway in operation, at least in our own country. The horse will then be once more returned to his legitimate field of labor, and the street-car passenger will be transported at an increased speed, and with all the comforts of easy riding, in cars propelled, lighted, and heated by electricity; while it is by no means improbable that, with further work on the line indicated, the passenger may step aboard a train in New York at ten in the morning, and eat a five-o'clock dinner in Chicago on the same day. Enough has indeed been accomplished to show that electricity is destined to be one of the most powerful factors entering into our social conditions, and that the ease of distribution and convenience of power afforded by it must bring forth changes in the social order which are even now hardly imagined.

ELECTRICITY IN LIGHTING

By HENRY MORTON.

Sir Humphry Davy's Production of an Electric Light in 1808—Ingenious Mechanism for Regulating the Carbons—Faraday's Discovery of Magneto-Electric Induction—Pixii's Machine, 1832—The Alliance Dynamo—Gramme's Armature Ring—The Construction of a Modern Armature—The Jablochkoff Candle—Other Devices in Arc Lights—Hard Carbon for Incandescent Lights—The Starr-King Lamp of 1845—Edison's Platinum Lamp—Experiments of Dr. Crookes with Platinum Wire in a Vacuum—Description of an Incandescent Lamp Factory—Systems for the Distribution of Electricity—The Series, Multiple-arc, and Three-wire Methods—Use of the Converter—Discoveries of Faure and Planté—Storage-Batteries.

IT was, we think, in reference to some electrical experiment that Benjamin Franklin made his often quoted and most suggestive answer to the question, What is the use of it? by another question, What is the use of a baby? and nothing has better illustrated the way in which scientific discoveries, like babies, can grow into usefulness than has electricity in its various developments and applications, among which by no means the least is that to electric lighting.

Indeed, this scientific infant, whose birthplace may be said to have been Sir Humphry Davy's lecture-room in the Royal Society, has not only developed into vigorous youth and useful manhood, but has also produced an extensive family of descendants, so wide-reaching and diverse in their characteristics that they must be discussed under numerous heads and various classifications, and have in many cases little in common with the founder of their family, except that electricity is the form of energy which vitalizes them, and that light is the result and evidence of their vitality.

Sir Humphry Davy in 1808 showed on a grand scale, with a galvanic battery of some two thousand pairs of plates, that when an electric circuit, established between two pieces of charcoal, was gradually interrupted by their separation, an arch or arc of dazzling light was developed between the separated pieces of carbon.

The magnificent intensity of this light attracted to it the attention of the world, and dreams as to its utility and applications were freely indulged in by many possessed of lively imaginations, but for many years there seemed little prospect that any of these dreams would be realized.

The radical and fatal difficulty was the cost of the electric energy required. Numerous improvements were made in the galvanic battery, by which its constancy of action and compactness as to bulk and weight were improved; but it always remained, and remains to-day, that the cheapest source of energy available in a galvanic battery is metallic zinc, and that metallic zinc is a costly material, with a low efficiency as compared with other substances, such as carbon or carbonaceous compounds, usually employed in the production of light. Left to the galvanic battery, therefore, the electric light, brilliant as were its capacities, would have been confined to the lecture-room of the professor and an occasional display in the theatre or opera-house, or out-of-doors on rare occasions, such as peace illuminations or national anniversaries.

In one direction much labor was spent and much improvement was made; that is, in the structure of "electric lamps," or "regulators," for the electric light.

When the electric arc is formed between the carbon terminals it causes them not only to glow and actually burn, but also to be vaporized and dissipated, so that they are consumed with considerable rapidity, and this, too, at an unequal rate, the positive terminal consuming much faster than the negative one. To provide for this, means of feeding the carbons (which for this purpose were made in the form of long, cylindrical rods of the most compact and refractory kinds of carbon, such as plumbago or gas-coke) toward each other as they were consumed must be provided.

Very ingenious and efficient "lamps" or regulators were constructed at an early date. There is one now in the cabinet of the Stevens Institute of Technology, Hoboken, N. J., which was imported some time prior to 1853, and used in some of my public lectures more than twenty-five years ago. It was designed by the eminent French physicist, Foucault, and constructed by the widely-known instrument-maker, Duboscq Soleil, of Paris.

Lamps similar in general principle, but different in their mode of operation, were made by Deleuil, Serrin, and Duboscq in France; by Roberts, Slater & Watson, Staite, and Chapman in England; and, indeed, as far as anything that could be done with galvanic batteries was concerned, there was nothing to be desired as regards perfection and efficiency in the electric lamp or regulator of the electric light.

This child of Sir Humphry Davy had reached his full growth and intelligence, and had attained not only a brilliant but a well-regulated manhood. His usefulness to the world at large, however, as I have already pointed out, was limited by the costliness of the apparatus by which his vital energy was supplied. Having thus, after the manner of the novelist, followed one of our characters up to a position of difficulty, we will turn in another direction and look after the other who is to relieve the situation.

Again we have the birth of a great scientific discovery, and this time it is in the laboratory of Michael Faraday at the Royal Institution.

Here magneto-electric induction first saw the light, and it was first demonstrated that an electric current could be produced without any galvanic or chemical action, by the mere motion of a conductor before a magnet.

The theory and detailed conditions of this action were fully

7

explained by Professor Brackett in the chapter on "Electricity in the Service of Man," and I will therefore say nothing of these, but pass at once to the practical application of this great discovery, which was soon made, and which, through a number of developments, has culminated in the dynamo-electric machine of to-day, which turns the mechanical energy of a steam-engine, of a waterfall, or of any other like motor, into an electric current, and thus enables us to secure electric energy from cheap and highly efficient coal or the like, instead of seeking it in costly and inefficient zinc.*

The first development of Faraday's discovery was made by Pixii, of Paris, who, in 1832, constructed an apparatus in which a large steel magnet was rotated so that its poles continuously and successively swept past those of an electro-magnet, or U-shaped bar of soft iron whose ends were surrounded with coils of copper wire.

This motion generated in the copper wire rapidly alternating electric currents, which were "commutated" or made to pass out of the machine in a constant direction by a simple "commutator" on the axis of the revolving magnet, which shifted the connections each time the direction of the current was changed.

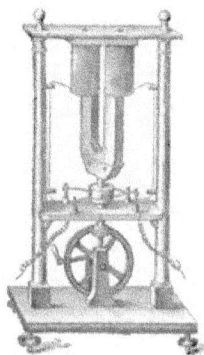

Fig. 1.—Pixii's Magneto-electric Machine, 1832.

The machine of Pixii is shown in the accompanying Figure 1.

In this, near the top, are seen the copper-wire coils wound on cores of soft iron like thread on a spool. Immediately below these is the permanent magnet, of a U-shape, and so supported that it can be rapidly rotated about a vertical axis midway between its poles, so that each pole is caused to approach, pass,

and recede from, in succession, each of the iron cores of the coils. Immediately below the bend of the U-magnet are the commutator

---

* The total efficiency of a pound of zinc is only one-sixth that of a pound of carbon.

segments, pressed upon by the contact brushes, and below these again is the gearing by which the magnet is made to rotate.

Machines operating on the same principle, but varying in construction (as, for example, by rotating the electro-magnet or coils of copper wire while the steel permanent magnet remained stationary) were brought out by Saxton, of Philadelphia, in 1833 ; by Clark, of London, in 1834 ; and by Page, of Washington, in 1835.

None of these machines, however, were of sufficient size to be available for the production of a practical electric light, although they all exhibited a capacity for this effect on a minute scale.

The first magneto-electric machine of a magnitude sufficient to operate a practical electric lamp was that produced by the united labors of M. Nollet, Professor of Physics at the Military School of Brussels. and his assistant constructor, Joseph Van Malderen, under the auspices of a corporation composed of French and English capitalists and known as the " Alliance Company."

Strange to say, this machine was built with the absurd object of using it to decompose water and employ the resulting gases in the production of light.

This machine, with some modifications by Mr. Holmes, of England, was, under the superintendence of Faraday himself, introduced into two of the English light-houses, *i.e.,* at South Foreland and at Dungeness. Its preliminary trial was made in 1857. The electric light was first thrown over the sea from the South Foreland on the evening of December 8, 1858, and from Dungeness on the 6th of June, 1862.

Figure 2 shows in outline one of the Alliance machines, as modified

Fig. 2.—An Alliance Dynamo used in the South Foreland Light-house, 1858.

by Mr. Holmes, which was long since put in operation at the South Foreland Light-house.

The outer framework supports twenty-four compound steel permanent magnets, and a drum inside carries thirty-two armatures or spools of copper wire wound on iron cores. As these pass from pole to pole between the magnets, currents are developed which are carried off by commutators on the farther end of the shaft, not shown.

The electric light was not introduced into the French light-houses until December 26, 1863, when it was installed at La Hève, near Havre. It was also used for lighting works of construction, such as the Cherbourg Docks, and on some vessels, for example, on the Lafayette and the Jerome Napoleon.

Although Faraday lived to see the little spark, which he had developed from a magnet and coil of wire in his laboratory, grow into these magnificent illuminators of sea and land, it was not until after many years and numerous new developments that the electric light approached the commercial utility which it to-day possesses.

These Alliance machines, on account of their great size and multitude of parts, were very expensive. Thus the two machines placed in the Dungeness Light-house, with their engines, appliances, and lamps or "regulators," cost £4.760, or nearly $24,000. The two located at Souter Point in like manner cost £7,000, or about $35,000, and the machines and accessories for the two lights at South Foreland cost £8.500, or about $42.500. The same characteristics caused them to be liable to accident and injury and costly in repairs. The world therefore waited for some further development before it could enjoy generally the advantages of electricity as a means of illumination.

The first of these came when Dr. Werner Siemens, of Berlin, constructed a machine in which the revolving coil or armature was made of the form shown in Figure 3, and was entirely enclosed between the ends of the permanent magnets. To construct this armature a long, solid cylinder of soft iron is taken, and two deep grooves are cut on opposite sides through its entire length, so that its cross section is such as appears at F in the accompanying figure. Insulated copper wire is then wound lengthwise in these grooves, its ends being united to the sections x, y, of the commutator. Journals on which this armature rotates are provided at either end, and at one end also a pulley by which it may be driven by a belt.

This armature secured a great concentration of action, by bringing the revolving armature into a highly concentrated field of magnetic force and allowing it to have a very rapid angular velocity of rotation. But the chief value of this improvement consisted in its serving as a step toward another, which was most remarkable in its results and excited the liveliest interest all over the world when it was announced. This next step was taken by Wilde, of Manchester.

He took a small magneto-electric machine, such as had been constructed by Siemens, and carried the current from its commutator to the coils of very large electro-magnets, which constituted the field-magnets of a similar machine, which, however, differed from the other, or Siemens machine, both in size and in having its field constructed of electro-magnets in place of permanent magnets.

Fig. 3.—Magneto-electric Machine of Dr. Werner Siemens.

Figure 4 shows such a combination, in which the first or small magneto-electric machine is mounted on the top of the other, and sends the current from its commutator through the coils of the electro-magnet below, between whose expanded poles another Siemens armature is made to revolve.

Fig. 4.—The Wilde Machine.

Small machine or feeder, with permanent magnets on top; large machine, with electro-magnets yielding available current below.

Under these circumstances the current developed in the armature of the upper machine by its permanent steel magnets will

develop a more than tenfold greater magnetic force in the poles of the electro-magnet of the lower machine ; and the second armature, rotating in this powerful magnetic field between the poles of this large electro-magnet, will develop a more than tenfold greater current than that of the smaller machine.

This method of multiplying or creating magnetic force was a wonderful discovery, and, combined with the use of electro-magnets in place of permanent magnets for the production of the magnetic field, gave an

Moses G. Farmer.

important increase in power and efficiency to the machine ; for, as compared with permanent magnets, the power of electro-magnets is vastly greater.

This advance, made by Wilde on April 13, 1866, was quickly followed by another, made almost simultaneously in Europe by Varley, Siemens, and Wheatstone, and nearly a year earlier in this country by Mr. M. G. Farmer, whose work in another department of electric lighting we shall have occasion to mention farther on.

This development may be indicated by the term "self-exciting," and consisted in the discovery that if the commutator is so connected with the coils constituting the field magnets that all or a part of the current developed in the armature will flow through these coils, then all permanent magnets may be dispensed with, and the machine will excite itself or charge its own field magnets without the aid of any charging or feeding machine such as the little one shown in Figure 4.

There is in all iron, unless special means have been taken to remove it, a little magnetic force. This small magnetic force, called "residual magnetism," in the iron cores of the field magnets will produce a little current in the armature when it is revolved. This current flowing through the coils of the field magnets will increase their magnetic force, and thus cause them to develop more current in the armature, which in turn, flowing through the coils of the field magnets, will further increase their magnetic force, and so on until a maximum, determined by the structural conditions of the machine and the amount of driving force applied to the pulley of the armature, is reached.

In practice such machines are each complete within themselves. When started they develop for a few moments only very feeble currents; but within a few seconds they "wake up" by degrees, and reach their maximum in less time than it takes to read this paragraph.

One other radical improvement in dynamo-electric machines remains to be recorded, namely, that due to the French inventor Gramme.

The essence of this lay in the structure of the armature. While previous to Gramme all armatures had been constructed either like spools of cotton or like balls of yarn wound on blocks, he made his armature by starting with an

Théophile Gramme.

iron ring (itself consisting of a coil of soft iron wire) and winding
the copper wire on this by passing the end of the wire again and
again through the ring. A Gramme
armature ring, cut and bent out partly,
and with some of its copper coils removed, is shown in Figure 5.

The cut ends of the iron wires constituting the ring-core are shown at A,
and B shows a portion of the copper-
wire coils wound around this ring-core.
The copper wire is continuous throughout as regards its electric connection,
but at frequent intervals a loop of this wire is carried out and attached
to a segment of the commutator.

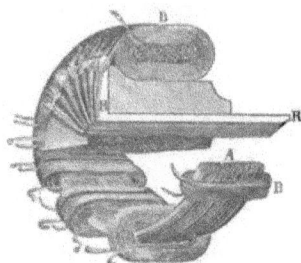

Fig. 5.—Section of a Gramme Armature
Ring, showing its construction.

This armature being rotated in a magnetic field (i.e., between the
poles of powerful field magnets) tends to deliver a substantially continuous current to "brushes" touching the commutator segments at
points midway between the poles of the field magnets.

It will be remembered that the iron ring constituting the core of
the Gramme armature was made of iron wires, and not of a solid piece
or ring of iron. The object of this was to prevent the formation of
electric currents in this ring-core itself, commonly called Foucault currents, which would be a cause of inconvenience by heating the armature and of loss by wasting energy in the useless production of this
heat.

The Siemens armature had no such provision, and accordingly very
serious difficulties were experienced in the running of machines using
such armatures, by reason of the intense heat there produced. Arrangements were in fact made in many machines to relieve this symptom
by running cold water through the armature, made hollow for that end;
but this did not cure the disease or prevent the loss of efficiency caused
by the conversion of the driving energy into useless heat in place of
useful current.

The desirable end was, however, soon secured by "laminating the armature core," that is, making it up out of a great number of thin

sheets of iron insulated from each other and held together by one or more bolts. The building up of such an armature core is illustrated in Figure 6.

The merit of this invention appears to have been assigned by the United States Patent

Fig. 6.—Workmen Building up the Armature Core of a Modern Dynamo.

Office to Mr. Edward Weston, of Newark, N. J., who on September 22, 1882, filed an application in the United States Patent Office describing such a laminated armature core, for which two patents were granted April 16, 1889, being Nos. 401.668 and 401,669.

We have given above all of the radical steps or improvements by which the dynamo-electric machine of to-day has been developed from the earlier constructions of Pixii, Clark, Saxton, and Page, or, in fact, from the experiment and discovery of Faraday.

There were, however, during the same time, a multitude of minor modifications of structure and arrangement introduced by various inventors, some useful and some useless, and when the world had been startled and interested by some of the wonderful developments, such as those of Wilde and of Gramme. it was found that in some forgotten patent or other publication some description might be read more or less completely anticipating these important discoveries.

We have not attempted to follow out the subject in this relation, which, however important in its legal consequences as effecting the rights of patentees, is not a part of the general history of the actual development of the electric light which we have attempted to write.

An endless variety has also been given to the forms and arrangements of the more recent dynamo-electric machines manufactured by the various companies, but these it would likewise be impossible for us to discuss within the limits of this chapter.

I will therefore select a typical case, and give some account of its mode of construction.

The most difficult and important part of the structure is the armature, and in building this the first thing is the laminated iron core. For this purpose an immense number of thin disks of sheet iron are cut out, each having a central hole to admit the shaft, and several other holes for the bolts which are to hold the series of disks together, so as to make of them a solid drum. These disks are then piled one upon another around the iron shaft which is to form the axle of the armature, as shown in Figure 6, and thick iron end-plates are applied at either end and bolted together by iron bolts going through from end to end. The drum or cylinder thus formed is then mounted in a lathe and turned to a smooth surface, except for such projections as may be left for guides in winding on the copper wires.

This is the next operation to be performed, and is shown in Figure 7,

Fig. 7.—Winding an Armature.

which represents the winding of a large armature intended to produce a very heavy current, and therefore wound with thick wire.

The workman in front is drawing the insulated copper wire down from a drum overhead and passing it lengthwise around the armature-core, which is supported by its axis in a lathe, while another workman assists him in pressing the wire accurately into place and keeping it close to the core. This wire is not wound on continuously, but in a number of short sections whose ends are seen sticking out somewhat irregularly. These ends are to be attached to the successive sections of the commutator, which is presently to be passed over the end of the shaft, appearing at the left.

Figure 8 shows just such an armature as that in Figure 7, but finished and turned the other way, so that, though the position of the observer is reversed, he still sees the commutator end of the armature

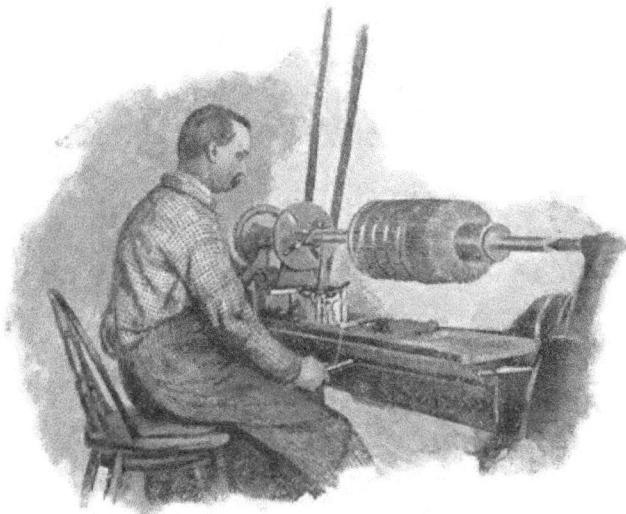

Fig. 8.—Finishing an Armature.

turned toward him. The numerous radiating lines at the nearer end of the drum are parts of the commutator-sections, which are attached at their outer ends to the successive coils of wire on the armature. At their nearer ends these radial bars bend at right angles, so as to pass along the surface of the shaft, being insulated from it and from each other by mica or other appropriate substance.

The workman in this figure is engaged in putting on the last turns of binding wire, which is wound in several bands, as shown, around the armature, not for any electric action, but to hold the coils, which run lengthwise round the drum, firmly in place and prevent them from being spread outward by centrifugal force when the armature is in use.

Fig. 9.—Field Magnets and Frame, without armature

These binding wires are made of German silver, a bad conductor for a metal, and are thoroughly insulated from the copper wires of the armature.

The armature. having been thus constructed, is now ready to be mounted in the framework of field magnets, which has been constructed in another department of the factory.

This is shown in Figure 9, and consists of a massive framework of cast-iron, portions of which are surrounded with coils of insulated copper wire so as to make the central parts of the upper and lower horizontal masses respectively north and south poles.

It is in the cylindrical hollow between these that the armature rotates, one end of its shaft being supported in the journal-box seen at the right, while the other end is supported in a journal-box out of view on the other side of the machine.

The adjustable supports to hold the brushes, or elastic strips of copper which press against the commutator and take off from it the current generated in the armature, are then attached to the sides of the bracket which carries the nearer end of the shaft, and the machine is substantially complete, the driving pulley being, of course, attached, when needed, on the farther end of the shaft.

Figure 10 shows this same machine completed in all respects, with

Fig. 10.—A Complete Dynamo—armature in place.

the armature inserted, the brushes in place, and the driving pulley on the farther end of the shaft.

The dynamo-electric machines of Weston, of Edison, of Brush, of Thomson-Houston, of Westinghouse, and a dozen others are all constructed (with considerable variations in form and detail) in the manner above described, and by their aid mechanical energy can be transformed into electric energy with an economy entirely unparalleled by any transformation heretofore known to the arts. Thus in the steam-engine we may, under very favorable conditions, transform ten per cent. of the

energy of the fuel into mechanical energy; but under the average working conditions we only secure about five per cent., the other ninety-five
per cent. being lost.

In the dynamo-electric machine, on the other hand, it is very common to secure a transformation of eighty per cent. of the mechanical
energy, applied to the driving pulley, into electric current, and in many
cases as much as ninety per cent. is so transformed and only ten per
cent. is lost.

Cheap electricity having been thus secured by the development of
the dynamo-electric machine, the electric regulator or lamp acquired a
new importance, and new demands were made upon the inventive
genius of the world on its account.

As long as expensive batteries were the only sources of electric
energy, it was considered quite enough to operate one lamp at a time;
but when the great capacities of the dynamo-machine were to be realized, it became clear that for economical working many lamps must be
operated from one machine, and, if possible, in a single circuit or one
after the other. For this the old regulators were not adapted. They
all operate in the following general method:

The current which supplies the lamp passes through an electro-
magnet, which controls a clock-work or other mechanism which allows
or causes the carbon poles to approach each other whenever the strength
of the current is reduced. As soon, therefore, as the burning away of
the carbon poles causes an increase in the resistance of the arc or space
between them by increasing its length, the resulting diminution of the
current causes the electro-magnet to release or actuate the feeding device, until the poles are brought near enough to diminish the resistance
of the arc to its normal amount.

With a single lamp in circuit this is all that is required, but it will
be manifest that anything which causes a diminution in the current
will cause the carbons to be brought nearer. Now suppose that two
such lamps are arranged in series so that the current flows first through

one and then through the other, and that, as must always be the case, one mechanism is a little (no matter how little) more sensitive than the other; then, when either pair of carbons burn away enough to diminish the total current to the point at which the more sensitive mechanism will act, that mechanism will so act, and will bring its carbons toward each other, until the resistance is diminished far enough to restore the normal current, and this will happen without the less sensitive mechanism being brought into action at all. This operation will then go on; the carbons of the less sensitive lamp burning away farther and farther, and their increase of resistance being made up by the approach of the carbons of the more sensitive lamp until the latter is extinguished by the actual contact of its carbon poles and the less sensitive lamp has secured an excessively long arc which is absorbing the entire energy of the circuit.

The same thing would happen with any number of such lamps in series. The most sensitive of the lamps would do the adjustment for all the rest, until its poles were brought into contact, and then the next in order of sensitiveness would take its turn, and thus one after the other would be thrown out of use, and the entire energy of the circuit would be concentrated in an abnormally long and probably destructive arc in the least sensitive lamp. Numerous plans were suggested to meet this difficulty, but the only ones which have reached any general practical success are those of Jablochkoff and of Brush.

Jablochkoff substituted for the lamps whose carbons were moved by mechanisms of some sort his electric candles with immovable carbons. In these the two carbon rods were placed side by side, vertically, very near to each other, the space between being filled with plaster-of-Paris.

An arc having been established between the upper ends of the carbons by a thin strip of carbon, which was quickly burned away, the same continued as the carbons consumed, because the plaster-of-Paris between them melted and volatilized as fast as the carbons were consumed. (Figure 11.)

These Jablochkoff candles were used to a considerable extent in

Europe in the early days of electric lighting, but never made much progress in the United States, being very inferior in efficiency and economy to lamps arranged on the Brush or other similar systems.

The arrangement first introduced in this country, as I believe, by the Brush Electric Co., and now universally used in one or another modification, may be described in general terms as follows: There are two electro-magnets or coils controlling the feeding mechanism which tend to oppose each other in the motions they produce.

Through one of these the current passes which also traverses the arc of the lamp, but the other magnet or coil is traversed by a current branching from the former where it enters the lamp, and rejoining it where it passes out, but not going through the arc.

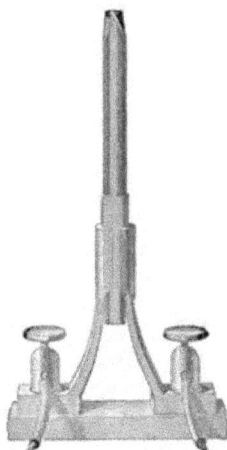

Fig. 11.—The Jab'ochkoff Candle.

This last-named coil has a higher resistance than the other, and normally transmits but a small fraction of the current as compared with that passing through the arc and the other coil.

If, now, by the burning away of the carbons, the resistance of that circuit is increased, two things happen at once: the current through the other coil, which is not in circuit with the arc, is increased at the same time that the current through the arc and *its* coil is diminished, so that the total current through the lamp remains substantially unchanged, and therefore nothing which happens in one lamp has any effect on the circuit at large or on any other lamp. Also the opposite magnetic effects in the two coils cause a rapid readjustment of the carbon electrodes and a consequent restoration of the arc to its normal length.

After this arrangement had been developed by the Brush Electric Co. some old patents were discovered in which the same principle was to a greater or less extent set forth; but, as in the case of the Pacinotti

8

article and the Gramme machine, these do not seem to have had anything to do with the practical development of the art of electric lighting prior to Mr. Brush's invention.

As with the dynamo-electric machines, so with the regulators or electric lamps for arc lights: their varieties of construction are endless, but they all come under the general description of holders for the carbon rods, whose motions are controlled by feeding mechanisms.

Fig. 12 — A Central Brush Dynamo Station.

which are in turn controlled by electro-magnets through which the operating current flows.

Such structures have reached a marvellous perfection as regards their regularity and certainty of action. Among the thousands of lamps which light our streets and stores night after night, a failure in operation is almost unknown to the ordinary observer. Irregularities, such as are incident to the unequal burning away of the carbon points, of course frequently occur; but the extinction of a light through any failure of the mechanism of the lamp is of the rarest occurrence even

where the lights are placed in the most exposed and inaccessible positions. A striking example
of this was furnished in the
lights erected and main-
tained for some time by our
Light - house Department
at Hallett's Point, N. Y.,
for the purpose of lighting
up the difficult channel of
the East River, known as
Hell Gate, illustrated in
Figure 13. These lights,
nine in number, arranged
so as to form about three-
fifths of a circle, were supported at a height of two
hundred and fifty feet by
a light iron tower. Each
light gave, by actual measurement, an amount of light
equal to three thousand
standard candles, or about
four times the light given
by the ordinary electric
lights employed in streets
and buildings.

These lights were put
in operation on October 20,
1884, and produced a mag-
nificent effect, lighting up
the whole surrounding town

Fig. 13.—Hell Gate Light, New York, before it was abandoned.

of Astoria and the adjacent channel. After several years of use it was,
however, decided that they did not afford the expected aid to naviga-

tion, and they were removed in 1888. During all these years, however, there was no failure caused by the mechanism of the lamps.

The number of arc lamps which are nightly operated by the different electric lighting companies in the city of New York is probably over five thousand, and throughout the United States it reaches 137,441.* Assuming that these lights are worth to their users the moderate rental of fifty cents a night, this represents an output of light having a value of $11,250,000 each year of three hundred days; all earned by this one branch of the family directly descending from the baby spark born from a magnet in the laboratory of Michael Faraday.

Admirable as is the system of electric-arc lighting for use in streets and open spaces, and in workshops or large halls, it is entirely unfit to take the place of the numerous lights of moderate intensity employed for general domestic illumination.

For this purpose it was at a very early period perceived that the incandescence or heating to luminosity of a continuous conductor by an electric current was the most promising method. It was also at a very early period perceived that the conductor to be used for this purpose must be one which would admit of being raised to a very high temperature without being melted or otherwise destroyed. The first material which was thought of in this connection was platinum, or one of its allied metals, such as iridium, which have the highest melting-points among such bodies, and are besides entirely unacted upon by the air at all temperatures. In 1848 W. E. Staite took out a patent for making electric lamps of iridium, or iridium alloys, shaped into an arch or horseshoe form.

One of the most serious difficulties, however, even with these materials, was that, to secure from them an efficient light, it was necessary to bring them so near to their fusing-points that a very minute increase in the current would carry the temperature beyond this and destroy the lamp by fusing the conductor. An escape from the

---

* From *Electrical Industries*, Chicago, August, 1890.

difficulty was offered by the use of hard carbon, such as that employed for the electrodes of arc lamps; but here the compensating drawback was encountered that this substance, when highly heated, was attacked by the oxygen of the air, or, in other words, burned. To meet this, plans were devised for the replacement of the consumed carbon conductor and for its protection from the air by enclosing it in a non-active gas or in a vacuum.

Thus in 1845 a patent was taken out in England by Augustus King, acting as agent for an American inventor named J. W. Starr, for an incandescent lamp, the important parts of which are represented in Figure 14.

Here a platinum wire is sealed through the top of a small glass chamber constituting the upper end of a barometer tube. This platinum wire carries at its lower end a clamp, which grasps a thin plate or rod of carbon, and also a non-conducting vertical rod or support, which helps to sustain another clamp, which grasps the lower end of the carbon strip and connects it by a wire with the mercury in the barometer tube below.

By passing a current through the platinum wire, and thence through the upper clamp, carbon strip, lower clamp, wire, and mercury, the carbon strip could be made incandescent, and was to a certain extent protected by the surrounding vacuum.

Fig. 14.—The Starr-King Incandescent Platinum Lamp, 1845.

Though this lamp produced a brilliant light, it proved in various respects unsatisfactory, and was abandoned after numerous trials.

Other inventors, as, for example, Konn, of St. Petersburg, continued to work with rods or pencils of hard carbon and achieved a limited success, but the irregularity and brittleness of the material seem to have been an insuperable objection and drawback, and the problem of commercial electric lighting by incandescent conductors yet remained without a solution.

This was the state of affairs even up to the fall of 1878, when, as is claimed, Mr. William E. Sawyer, in combination with Mr. Albon Man, after many preliminary experiments, produced their first successful incandescent lamp with an arch-shaped conductor made of carbonized paper. In their application for a patent, filed January 8, 1880, these inventors use the following remarkable language in their fourth claim: "An incandescing arc of carbonized fibrous or textile material." This indicates that they realized the importance of what seem to be the common features of the present electric incandescent lamps, namely, the arc or arch or bow or loop form, and the carbonized fibrous or textile material. They also specially refer to carbon incandescent conductors made from paper.

After a long and hotly contested interference, the United States Patent Office has granted them a patent in which these points are broadly stated.

The lamp brought out by Messrs. Sawyer and Man, soon after their application for a patent, and described and shown in that application, was a rather large and complicated structure; and had no improvement and simplification of this structure been made, the present immense development in electric lighting would no doubt have been unattained.

It is to Mr. T. A. Edison, without doubt, that we owe many of the simplifications and modifications which, by cheapening the lamp and diminishing its weight, have extended its range of use and its usefulness to a remarkable degree.

On his return in the fall of 1878 from the far West, where he had gone in company with Dr. and Mrs. Henry Draper, Dr. George F Barker, and the present writer, to observe the total solar eclipse of that year, Mr. Edison visited the shops and laboratory of Mr. William Wallace, at Ansonia, Conn., where many experiments with electric-arc lights and dynamo-machines were in progress, and while studying these, was impressed with the desirability of producing an incandescent electric lamp.

Like so many before him, he first turned to platinum and platinum alloys, and devised a form of lamp admirable for its simplicity, but, unfortunately, open to a fatal objection. This first lamp of Edison's is shown in Figure 15, in which $a\ b$ is the incandescent platinum wire.

The announcement of a new system of electric lighting, made by Mr. Edison and his friends on the foundation of this device, attracted universal attention, and even caused a serious fall in the value of "gas stocks" in this country and abroad. It is, indeed, amusing now to look back upon the extravagant assertions and predictions made at that time, and widely circulated, when we realize how more

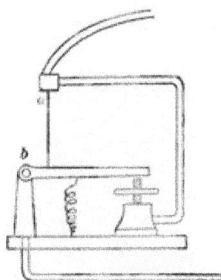

Fig. 15.—Edison's First Incandescent Platinum Lamp.

than frail was their foundation. In fact, Mr. Edison very soon found out that this simple device was entirely insufficient for the purpose proposed, because the heated platinum wire gradually stretched by its own weight, and thus was constantly getting out of adjustment, and finally would become attenuated and break.

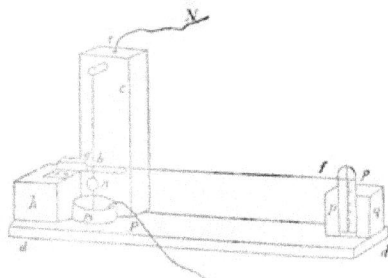

Fig. 16.—Dr. J. W. Draper's Plan for an Incandescent Platinum Lamp, 1847.

$a\ b$ is the incandescent platinum wire or strip, supported above by the brass pin at a, which runs into a cavity, e, filled with mercury and is thus connected with the battery wire N. The other end of the platinum wire or strip is attached to a delicate lever, $q\ p$, turning about the fulcrum g; a little weight, n, tends to keep the wire stretched, and communication is made through it by a copper wire with the cup m filled with mercury, into which dips the other battery wire P.

It also happened that, though the secret of this great invention was carefully guarded, some inkling of it escaped, and this enabled those who were familiar with such subjects to perceive the close similarity between this Edison lamp and a similar device constructed and used by Dr. J. W. Draper prior to 1847, and described and figured in articles published by him during that year in *The American Journal of Science and Arts, The Lon-*

don, Edinburgh, and Dublin Philosophical Magazine, and Harper's New Monthly Magazine. This apparatus of Dr. Draper is shown in outline in Figure 16. It was used by Dr. Draper as a source of light or lamp with which he determined the relations between temperature and luminosity. At the conclusion of his article Dr. Draper says: "An ingenious artist would have very little difficulty, by taking advantage of the movements of the lever, in making a self-acting apparatus in which the platinum should be maintained at a uniform temperature notwithstanding any change taking place in the voltaic current."

It also appeared that precisely the same idea had occurred to another inventor, Mr. Hiram S. Maxim, who has recently developed such a marvellous improvement in magazine or repeating guns, and who, on December 22, 1879, filed an application for a patent which, after an interference litigation with Edison, was finally issued to Maxim on September 20, 1881, for the form of electric lamp shown in Figure 17.

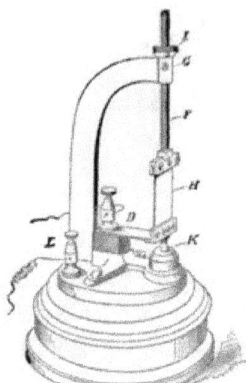

Fig. 17.—Maxim's Incandescent Platinum Lamp.

H is a strip of platinum, adjustably supported by means of the screw *F* and nut *I*, from the standard *G L*, from which it also receives an electric current which normally passes out through *D*; but when, by excessive heat, the platinum strip is elongated unduly, a short circuit is closed at *K*, which diverts the current from the standard and platinum strip, and so prevents the fusion of the latter.

It has also been shown that in 1858 Mr. M. G. Farmer, one of the veteran electricians of America, to whose work in connection with the dynamo-electric machine allusion has been made before, lighted a room in his house at Salem, Mass., for several months, with platinum lamps of similar structure controlled by automatic regulators.

During 1878 and 1879, however, Mr. Edison was most diligently at work, and, perceiving the imperfections of his first ideas, sought in every way to overcome them. It thus came to pass that by December

21, 1879, at which date he made his first rev-
elation to the public, in the pages of the New
York *Herald*, he had perfected a platinum lamp
which is shown in outline in Figure 18, as well
as some other forms substantially like it.

But these platinum conductor lamps were
not the only outcome of Mr. Edison's work
between the fall of 1878 and December, 1879.
As this *Herald* article also related, Mr. Edison,
like many before him, having experienced the
insuperable difficulties present in metallic conductors, had turned his attention to carbon in
various forms; and, like Sawyer and Man,
had found fibrous textile materials, when carbonized, to be most convenient, and paper especially to be, in the first instance, the most
available substance. Like Sawyer and Man,
he had also found the arch, or horseshoe form,
to be the most desirable.

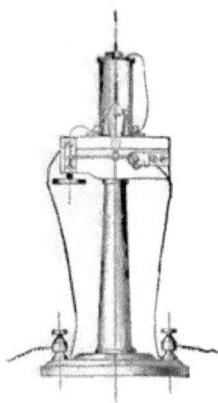

Fig. 18.—Edison's Platinum
Lamp on Column Support, 1879.

The incandescent wire of a
platinum alloy is supported by a
metallic rod about which it is
wound and whose expansion
serves to operate a shunting contact below, by which an inconveniently high temperature is
avoided.

Though working with the same materials and form, Edison produced a structure very different in appearance from that of Sawyer and
Man, as will be seen by reference to Figure 19, which represents one of

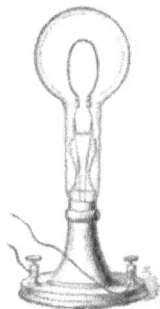

Fig. 19.—Edison's Paper
Carbon Lamp.

Edison's paper carbon lamps, which was the first one
whose electric properties were accurately measured,
these measurements having been made at the Stevens
Institute of Technology, early in 1880, by the present
writer, acting in his capacity as Chairman of the Committee on Scientific Tests of the United States Light-
house Board, that body desiring information as to
this new light, and deputing the work of investigation
to this committee.

In this lamp the carbon conductor is supported
on platinum wires and held in minute platinum

clamps at the ends of these wires, which are sealed through the walls
of the pear-shaped enclosing tube in the manner which had been familiar for twenty years in the construction of the beautiful electric toys
known as " Geissler tubes."

The interior of this glass vessel had likewise been exhausted and

Dr. William Crookes, F.R.S.

hermetically sealed in the manner usual with many Geissler tubes and
with the radiometers of Dr. William Crookes.

Indeed, as was subsequently made apparent, the wonderful results
obtained by Dr. Crookes in the production of very perfect vacua were
of essential importance to the development of the incandescent electric

lamp. Several of the instruments produced by Dr. Crookes in the course of his researches were in fact incandescent electric lamps, consisting of coils of platinum wire enclosed in glass vessels exhausted to a very high degree, the coils being heated to brilliant luminosity by electric currents. One of these is shown in his paper in the " Philosophical Transactions for 1876," vol. xlvi., Part II., page 351.

Further experience proved to Edison and others that *paper* carbons were not the best for the conductors of electric lamps, and many other substances have been, or are now, employed for this purpose. Among these may be mentioned silk, hair, parchmentized cotton thread, tamodine or reduced celluloid, and last, but not least, bamboo, which is used to a very large extent.

The making of these electric lamps is carried on in a number of large factories, such as that of the Edison Co., at Harrison, near Newark; those of the Westinghouse Electric Co., at Newark and at Pittsburgh; that of the Consolidated Electric Co., at West Twenty-third Street, New York; that of the Thomson-Houston Co., at Lynn, Mass.; that of the Brush Co., at Cleveland, O., and a number of smaller establishments elsewhere. The daily output of all these factories taken together is about fifteen thousand lamps, or four and a half million a year.

The methods of manufacture are substantially alike in all, and I will therefore describe one only as an example.

Sheets of tamodine (or celluloid from which the nitric constituent has been removed) are cut by a machine into delicate strips or filaments, which are collected in small bundles and bent so as to lie in U-shaped grooves in iron plates. These, packed with carbon powder, are enclosed in large black-lead crucibles, carefully closed, and heated in a Siemens furnace to an intense white heat. After cooling, the crucibles are opened, and the now carbonized filaments, looking like delicate wires or threads of steel, are removed. They have now the U-shape into which they were bent before carbonizing, but are so elastic that they can be stretched out straight without breaking. Their ends are next thickened by a remarkable process devised by Messrs. Sawyer and Man, and which is conducted as follows: Each U shaped fibre is grasped by two clamps, one holding it by the extremities or ends, and the other at a little distance above. The loop and clamps are then plunged in a vessel of high-boiling petroleum-oil, like the well-known "astral oil," and a powerful electric current is passed from the clamps through the short portions of the filament, near its ends, which are grasped between them.

By this means these portions are intensely heated and decompose the hydrocarbon liquid in contact with them, so as to plate themselves with compact carbon like that deposited from the gas in the necks of gas-retorts. A few seconds' action suffices to make this deposit of carbon thick enough to answer the desired purpose.

We will next turn to the glass-blowing department, where hundreds of girls are employed in all the delicate and skilful manipulations involved in the glass-work of these lamps.

The first step is to take two minute pieces of platinum wire, one end of each having been shaped into a little socket capable of holding the enlarged end of the carbon filament; and, after mounting them in a small lathe-chuck, to wind melted glass from a glass rod, heated in a glass-blower's lamp, around these platinum wires until they are for some distance embedded in glass and formed into a structure such as is seen at the lower part of the ordinary incandescent lamps. Into these glass and platinum supports are then inserted the enlarged ends of the carbon filaments.

In the meantime small glass flasks, made by the thousand at the glass-works, are passed through a variety of manipulations by which a small glass tube is attached to what would be the bottom of each flask, and its neck is shaped so as to receive the glass socket carrying the platinum wires and carbon filament. At the proper time this socket is dropped into the prepared flask, and by manipulation with the glass-blower's lamp and a sleight of hand which is simply marvellous, the glass socket, with its carbon filament and connecting wires, is sealed, by fusion of the glass itself, into the neck of the flask.

This operation is shown in progress in Figure 20, where the girl in the foreground holds in her left hand the glass flask by the glass tube

Fig. 20.—Sealing the Glass Socket and Carbon Filament into the Flask of an Incandescent Lamp.

which has been attached to it, and in her right hand the shears with which she at times holds and shapes the glass socket and neck of the flask. The blow-pipe flames, constituting what is called the "glass-blower's lamp" or "fire," are seen as pointed tongues of light between the hands of the operator, who is supposed at the instant represented to have just raised an electric lamp, finished (so far as her work is concerned), from the flame.

The next thing to be done with the lamps is to exhaust them. For

this purpose they are attached by the small glass tubes before mentioned to radiating glass connectors, and these are in turn attached to
the pumps, while at the same time electric connections are made so
that currents can be sent through the filaments of the lamps while they

Fig. 21.—The Process of Exhausting the Air from Incandescent Lamps.

are being exhausted by the pumps. These pumps are themselves
entirely composed of glass. and operated by the flow of mercury back
and forth within them, and their operation is so nearly automatic that
a few attendants can keep a large number of them in steady operation.

Fig. 22.—The Hoosac Tunnel Lighted by Glow Lamps.

When a good vacuum has been reached, the current is passed through the lamps and they are then kept at a brilliant incandescence for some hours, in order to drive out any gas which might be occluded in the carbon filaments or adhere to the interior surface of the glass. This process of exhaustion and a series of pumps and lamps in operation during the process are shown in Figure 21.

After the complete exhaustion of the lamps it then only remains to "seal them off," that is, to melt the small glass tube attached to each so that its sides close together, and it becomes a little knob of glass, and to attach the brass caps by which they are to be subsequently connected to their sockets.

The uses of these lamps are so countless and so familiar to every one that we have only selected one unusual one for illustration, namely, the lighting of the Hoosac Tunnel, which has recently been carried out by this means in the face of great difficulties encountered in securing adequate insulation, in such a situation, for the wires carrying the current to the lamps. The lamps are attached to the rock or to the stone lining of the tunnel in the manner shown in Figure 23, and produce when in operation the effect shown in Figure 22.

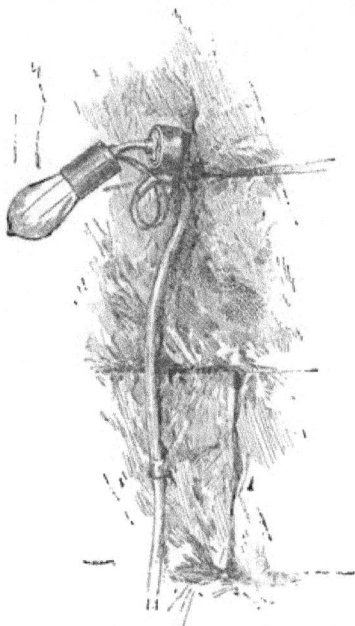

As we have seen so often already, the solution of one problem always opens up another, and thus it is not surprising that the cheapening of electricity and increased efficiency of incandescent lamps brought to the front the problem of an economical method for carrying the electric current from the generator to the lamps.

There were two well-known systems which had been often used in other applications of electricity, and, indeed, even described and patented for use in electric lighting, namely, what are commonly known as the "series" and the "parallel" systems.

Fig. 23.—Method of Attaching Glow Lamps to the Walls of the Hoosac Tunnel.

The "series" system is that always and necessarily employed whenever more than one arc-light is used on the same circuit, and may be likened to the arrangement of disks on the chain of a chain-pump, or illustrated by the accompanying diagram, in which X represents a

dynamo-machine and o, o, o, o, etc., represents a series of lights connected by the circuit wires —, —, so as to form a single chain from the

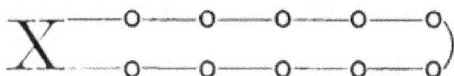

machine through all the lights in succession back to the machine
again.

This was the usual arrangement of the telegraph instruments at the
various stations on a line.

The "parallel" or "multiple-arc" system was one which might be
indicated by a ladder or by the accompanying diagram, where, as

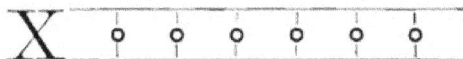

before, X represents the dynamo, from whose poles proceed two main
conductors, between which the lamps o, o, o, etc., are placed in cross
connections.

This was a method commonly employed in central telegraph offices
for operating the sounders by means of the large "local" battery. It
is also described in the United States Patent to H. Woodward for
improvement in electric lights, granted August 29, 1876, as well as in
many other places.

The first method has certain drawbacks which are specially important in the case of incandescent lamps, where, for economy, a large
number should generally be operated on a single circuit:

1. The extinction of one lamp means the extinction of all, unless
some more or less complicated mechanism is provided to restore the
connection around the lamp which has failed or has been turned out.

2. The electro-motive force, or electric pressure, needing to be multiplied in direct proportion to the number of lamps in the circuit, soon
becomes inconveniently high.

Both of these difficulties being avoided in the "parallel" system,
this last has been generally adopted by all the companies using incan-

9

descent electric lights for most of their work. This is, however, by no means universal, for the Edison Co., in what they call their "Municipal system" (used mostly for street lamps in small towns and villages), run incandescent lamps in series. Other companies often run their lamps in a combination of the two systems, and the Heissler Co. run their lamps in "series" exclusively.

In avoiding the difficulties of the "series" system mentioned above, the parallel or multiple-arc system encountered others, the chief of which was the great size and cost of the conducting wires, if the distance between the dynamo and the lamps was considerable. Suppose that a group of lamps was placed one thousand feet from a dynamo, and the wires used were of such a size that their resistance to the flow of the current caused them to waste ten per cent. of the energy developed. Now let us suppose that this group of lamps is moved away one thousand feet farther. This would, of course, mean doubling the length of the wires, which alone would double their cost; but it would also mean doubling their resistance, if they were not made larger than before, and so wasting twenty per cent. of the electric energy generated by the dynamo.

To avoid this loss we must make the wires twice as heavy per running foot, and if we do so we can then reduce the loss at two thousand feet to ten per cent. as before, but clearly we have four times the weight of copper to pay for in our conductors. If the lamps are removed to a total distance of three thousand feet, we shall have three times the length of wire, and to keep down its resistance to that producing a loss of only ten per cent., we must make the wire three times as heavy per foot, or, in all, we shall require nine times as many pounds of wire to operate the lights at a threefold distance. The law evidently is, that the weight and cost of the wire will increase as the square of the distance.

This difficulty is mitigated to a considerable degree by what is known as the "three-wire" system, first indicated by Mr. Brush in his patent No. 261,077, issued July 11, 1882, and developed in two different directions by Mr. Edison, in his patent No. 274,290, issued March 20, 1883, and by Mr. H. M. Byllesby, of the Westinghouse Co., in his patent No. 345,212, issued July 6, 1886, so that the loss can be reduced to three-eighths or even to five-sixteenths of what it would otherwise be by a moderate increase in the complication of the arrangements.

The outstanding loss has, however, led to the development of a radically new and very interesting system, known as the secondary or

Fig. 24.—An Alternating Current Dynamo and Exciter.

transformer system, chiefly represented in this country by the Westinghouse Electric Co.

The principle on which this system operates is indicated by Professor Brackett in the introductory chapter, and may be briefly stated by saying that if we have two conducting wires parallel to each other, and pass an interrupted or reversed (*i.e.*, alternating) current through one of them, there will be produced a similar, but always alternating, current through the other, without there being any conducting contact at all between the wires.

This may be very beautifully shown by the following experiment:

We have upon a table an oval coil of fine copper insulated wire, through which is passing the rapidly reversing or alternating current obtained from a dynamo-machine which is working without a commutator. (Figure 24.)

If, now, we hold above it just such another coil, in whose circuit is included an incandescent lamp, this lamp will light up and glow to its full intensity as we bring the second coil near to the first, and will die out as the coil is moved away. This will operate just as well with a plate of glass between the two coils.

This action is greatly intensified by enclosing both coils in a mass of iron, whereby magnetic influence is brought into play, and accordingly the converters or transformers used in this system are made, as will appear from inspection of Figure 25, by enclosing the two coils in a mass of iron made up of thin sheets, so cut that they can be sprung in, one at a time, around the coils.

The relative character of the currents in the two coils depends substantially on their lengths and consequent resistance; that which is shorter and thicker having a current of larger volume and less pressure or electro-motive force, and that which is longer and thinner having less quantity or current strength and more electro-motive force or pressure. Now, a current of high electro-motive force and small quantity can be carried a long distance on a small wire with very little loss.

Fig. 25.—Converter or Transformer Used with an Alternating Current.

If, then, we pass this current through a coil of long fine wire, in a converter whose other coil is relatively short and thick, we shall obtain in the latter a current whose quantity is great and whose electro-motive force is low. In other words, we can thus transmit such a current as goes easily on a small wire from the central station to the house where the lights are to be used, and there transform it into the kind of current most desirable for the operation of incandescent lights. In practice the Westinghouse Co. send out their currents with an electric pressure of one thousand volts or units of electro-motive force. A quantity of this current equal to one ampère, or unit of current strength, running through the fine wire of one of their converters will develop in the coarse wire a current of twenty ampères quantity but of only fifty volts pressure.

Such a current, however, would be just what was wanted to run twenty incandescent lamps in "parallel" series, which is the most convenient way, as each is then entirely independent of all the others.

The problems of cheap production of electric energy, of cheap and efficient regulators or arc lamps, of cheap and efficient incandescent lamps, and of economical methods of distributing the electric energy from the electric generators to the lamps having been solved so thoroughly, as has been here indicated, there seemed little yet to be desired. One thing, however, was *not* provided for, and that was the storage or accumulation of electric energy. The method of its production by the dynamo requires an absolutely constant activity and a literally sleepless vigilance. If the steam-engine stops or relaxes its speed, the light goes out or becomes dim; or if a belt breaks or slips off, or any part of the dynamo becomes disarranged, the light is gone in an instant and without warning.

This lack of storage capacity was often referred to, and was a serious reproach to the systems of electric lighting as compared with other methods of illumination. This reproach has been to some degree removed by the labors of M. Camille A. Faure, and of those who have followed up, and to a greater or less extent improved upon, his invention.

The "state of the art," as regards the storage of electricity prior to Faure, may be fairly expressed and summarized by a statement of what was done by Gaston Planté * in 1860.

* Planté died in June, 1889.

This experimenter took a series of lead plates, immersed in a vessel containing diluted sulphuric acid, and coupled or joined them so that they were united into two groups, each alternate plate constituting one group and the intermediate plates being connected so as to form the other group. He then passed the current from a couple of battery cells, arranged in series, into this structure, by joining the positive pole of the battery to one of these groups and the negative pole to the

Camille A. Faure—inventor of a storage-battery system.

other. When the action of the battery had continued for a long time, he found that on removing the battery he could get an electric current from his two groups of lead plates; this current being opposite in direction to that developed by the battery and capable of yielding a greater flow for a shorter time. The knowledge already accumulated had explained the cause of this, which was as follows: The plates of lead, even before immersion, were coated with a film of oxide, and on immersion, at all events, would soon acquire a coating of sulphate of lead. The passage of the battery current between these plates would convert the oxide or sulphate, on one side into metallic lead, and on the other side into peroxide of lead.

Now, metallic lead and peroxide of lead, as was well known, are substances well fitted to develop a galvanic current in the same way that such a current is developed by an ordinary galvanic battery made with plates, for example, of zinc and copper—the metallic lead taking the place of the zinc. There was, however, one important difference, that whereas in the zinc battery the zinc went into solution, in the lead battery nothing was dissolved, and therefore everything kept its original position, so that the original cycle of action could be indefinitely repeated. Planté, in fact, found that by repeatedly charging his lead plates from an ordinary battery, and discharging them again, and also by reversing the direction of the charging current, the capacity of his lead plates, or the amount of electric energy which they could be made to absorb and redevelop, was greatly increased. Indeed, the maximum capacity secured by this treatment was only reached after about six months of such charging and discharging. The reason of this also was not far to seek. By these repeated actions the surfaces of the leaden plates was corroded or honey-combed, and thus a greater amount of the material was in condition to be converted into metallic lead and peroxide by the battery current, and again to return to protoxide and sulphate during the discharge.

To obtain any considerable capacity in this way, however, required months of treatment (called "forming"), and a heavy expense for the charging currents, and soon after a battery was fully formed it began to deteriorate by a continuance of this corrosive action, which caused the porous material to scale off and the plates themselves to break up.

Planté's batteries were therefore of no commercial value, on account of their high cost and limited capacity.

Matters stood thus when, in 1881. the world was astonished by the accounts of what Mr. Faure had done in the way of improving this

Planté secondary battery into his electrical accumulator or storage battery.

His plan was a very simple one, but wonderfully effective. He took a quantity of litharge or of red lead, or a mixture of the two, both being oxides of lead, and making this into a paste with dilute sulphuric acid, he coated the lead plates with this mixture. When the plates so coated were plunged in dilute sulphuric acid, and an electric current was

Fig. 26.—Room in the Factory of the Electrical Accumulator Company.
The groups of battery plates, such as are shown hanging from the travelling pulleys, are immersed in tanks of dilute acid, and charged by electric currents.

made to pass between them, the thick coating of oxide paste on one side began at once to be converted into a spongy mass of metallic lead, and on the other into a like spongy mass of peroxide of lead.

In this way no time was lost in the "forming" process, and the capacity of the plates was very much greater in proportion to their weight than in the most perfectly formed plates of Planté. An improvement on this plan was made by Swan, of England, and others, which consisted in so perforating the plates that the paste of oxide would fill the apertures, like a series of rivets with conical heads, by which it would hold itself in position.

The Faure and Swan patents and some others were taken out in this country by the Electrical Accumulator Co., who established a large factory at Newark, N. J., where these batteries have been made for many years. Figure 26 shows the interior of the principal work-room in this factory. These batteries only *store electricity* in a metaphorical sense. What they actually do is to transform the active energy of an electric current into the potential chemical energy of separated chemical substances, which are able, by their reunion, to develop again an electric current such as that which produced them. In other words, the charging current each time decomposes the oxides and sulphates of lead formed by the chemical action of the battery during its discharge, so as to develop metallic lead on one set of plates and peroxide on the other. This having been done, this metallic lead, by combining with oxygen and sulphuric acid on the one hand, and the peroxide, by combining with hydrogen on the other, develop an electric current, as does any ordinary galvanic battery.

As these successive changes can be repeated an indefinite number of times, the effect and appearance are the same as if the electric current had been in fact stored up or accumulated in the storage-battery.

THE TELEGRAPH OF TO-DAY.

By CHARLES L. BUCKINGHAM.

THE BEGINNINGS OF THE ELECTRIC TELEGRAPH—SIGNIFICANT DISCOVERIES FROM DAVY TO HENRY—IMPORTANCE OF THE MORSE ALPHABET—EFFORTS TO INCREASE THE CARRYING CAPACITY OF THE WIRE—FARMER'S MULTIPLE-SYNCHRONOUS SYSTEM— THE DUPLEX REDUCED TO PRACTICAL FORM IN 1872—EXPLANATIONS OF ITS PRINCIPLES—THE DIPLEX AND QUADRUPLEX—THEORY OF THE MULTIPLE-HARMONIC—BAIN'S CHEMICAL AUTOMATIC TELEGRAPH—WHY AN AUTOMATIC SYSTEM CANNOT BE MORE THAN AN AUXILIARY—FAC-SIMILE TELEGRAPHY—TYPE-PRINTING MACHINES—THE SIPHON RECORDER FOR SUBMARINE CABLES—LOCATING BREAKS IN A CABLE—TELEGRAPHING FROM A MOVING TRAIN—STATIC RETARDATION.

THE successful projects of Professor Morse have a historical standing of nearly half a century. Meanwhile his work, overgrown with improvements, has become a commercial industry reaching to the outskirts of civilization. In the United States alone the Western Union Telegraph Company, with its 600,000 miles of wire, transmits annually more than 50,000,000 messages. According to the standard of Darwin, who assumes the making of fire to be the greatest discovery of man, an invention acquires its rank from the extent of its subsequent application. This striking advance from Morse's humble beginning may well lead our generation to inquire from what achievements his name has almost come to be a synonym for the telegraph.

In October, 1832, when his attention was first drawn to this subject, and even before he had so much as assumed the possibility of electrical communication, science had placed at his disposal the three essential

Main Operating-room of the Western Union, New York.—[Before the Fire of July, 1890.]

(Showing front view of switchboard ; the pneumatic system for transmitting messages to and from city stations ; and the mechanical system for collecting from and distributing to the 600 operators in the room.)

elements: a metallic conductor for conveying the fluid between distant points, a galvanic battery affording an ample source of electricity, and an electro-magnet for translating electric currents into intelligible signals. Following the discovery of the voltaic pile in 1800, Davy, before 1810, had employed the combined action of two thousand battery cells in experimenting with the electric light, and had developed currents stronger than would operate the longest telegraph-circuit of the present day.

In 1819 Oersted had observed that an electric current caused the deflection of the compass-needle, and in the year following Arago succeeded in magnetizing a steel needle by placing it across a wire conveying a current. Ampère immediately perceived the multiplied effect that would be obtained by coiling the wire around the needle, and in 1825 Sturgeon substituted for steel a core of soft iron. The electro-magnet, although crude in form, was then complete as an invention. In 1828, however, it was taken up by Professor Henry, and in his hands, before 1831, was advanced so far from a laboratory experiment that doubtless it could have been advantageously used as a telegraph-receiver. [See "The Electric Motor and its Applications."]

That Henry, during this period, placed the world in full possession of a knowledge of the character and properties of the electro-magnet cannot be doubted when we remember that he constructed a specimen, existing to-day, capable of attracting an armature to its poles with a force of more than two thousand pounds; and in 1831 he went farther and employed an electro-magnet in an experimental telegraph, which by vibrating a bell-hammer, audibly announced signals by the closing and breaking of the current. Whatever merit, therefore, there may be in the claim advocated for Professor Henry that he invented the telegraph before Morse, there is little room for doubt that he brought the electro-magnet to a stage of development fitting it to many uses for which it has since been discovered to be suited.

If, in 1832, Morse had appreciated the possibility of manually closing and opening a circuit to effect transmission, and of reading sound-signals produced by the blows of an electro-magnet's armature, he might, with little trouble and expense, have organized a telegraph system from the galvanic battery and the Henry magnet. But instead of forming a system of those parts, he adopted them as a skeleton upon which he built, not thinking that one day his additions would become obsolete and that the system would be brought back to the simple elements with which he began. He assumed that an automatic mechanism must be employed to insure accuracy of transmission, and that messages must be permanently recorded upon paper or other fabric; and to meet these requirements, whether real or imaginary, consisted in large part the work of introducing the electric telegraph. The first telegraph contrived by Morse reveals complications which are entirely omitted in systems where signals are read by sound. The devices added by Morse contained designs requiring the most delicate workmanship, and every part of the mechanism became a source of difficulty, threatening the entire undertaking with failure. The possible electrical obstacles to its success seem almost to have been forgotten, for those of a purely mechanical character were much more serious. It was not merely a question whether, electrically, the system was possible, but chiefly whether a rather difficult electrical experiment could survive the encumbering intricacies of the apparatus. But Morse's early plans, involved as they were, contained the groundwork upon which the dot and dash alphabet was produced by a natural evolution; and, whether his system was the best or poorest of its kind, it brought the telegraph to the favorable notice of capitalists in a form which could not fail, even in the hands of unskilled operatives.

It is said that Morse was chagrined that operators, as they became skilful, could read messages by sound without the aid of his permanent recorder; but, with respect to the credit due him, it matters not whether his devices had their uses for a year or for a century; they served their purpose and gave the telegraph an introduction to the world, which otherwise it might not have received for a generation.

If the struggles of Morse and his associates in securing public recognition of their undertaking could be forgotten, it certainly would now seem anomalous that he should be honored· by having his name metonymically represent the modern electro-magnetic telegraph, consisting as it does of a circuit, a circuit-breaker, a battery, and an electro-magnet—for these are the elements which were old, and to which he had recourse when he first assumed the rôle of inventor.

Others before him had devised systems of great merit, while many of his contemporaries, of higher scientific attainments, were diligently working in the same direction ; nevertheless, his success in adapting the telegraph to the ignorance of the age rightfully placed him beyond competition.

Doubtless Morse derived valuable assistance from Henry and Vail, but the telegraph of to-day bears the marks of his genius in features, from the smallest detail to things of indispensable importance.

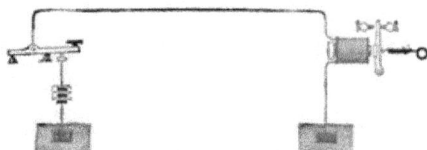

The world has lost noth-ing, nor is it less to his credit if parts of the invention which he esteemed most have, like the false works of an arch, been removed. When they became an incumbrance their absence was as important as had been their presence, to give the structure its original shape and strength.

The Modern Morse Telegraph.

No sooner had Morse and Vail demonstrated the feasibility of the telegraph than it became important to increase the carrying capacity of the wires. In 1846 Bain proposed to employ perforated strips of paper to effect automatic transmission in connection with an electro-chemical process for recording, in which marks upon a moving band of ˙paper are made by discoloration attending the passage through it of signalling currents. But up to 1852 no one appears to have conceived the possibility of a system by which two or more operators might simultaneously use a wire to transmit independent messages.

In that year, however, Moses G. Farmer, of Salem, Mass., devised a synchronous-multiple telegraph, in which he proposed to employ two rotating switches, one at each end of the line, to successively and simul-

Underside of the Switchboard for 2,000 Wires; Western Union Building, New York.—[Before the Fire.]
(Above, the wires as seen in the room beneath the Switchboard.)

taneously join the several operators at one station with those at another. For illustration, it may be assumed that at each end of the line an equal number of short wires is connected from the earth with a circular series of stationary electrical contacts arranged like the hour-marks of a clock-dial, over which a rotating arm, like the hand of a clock, rapidly draws a spring or trailing conductor. The rotating arms are connected with the main line, one at each end, while each of the short wires is provided with a set of Morse instruments, and thus it is that each operator may send a signal to, or receive one from, the operator upon a corresponding branch at the distant station. It is now seen that if the two arms are rotated together, having been started from the same angular position, the main line will simultaneously join the No. 1 branches at each station, and all the several branches at one end will, in rapid succession, be connected with corresponding branches at the other. When two branches are thus joined, a momentary electrical connection is made between the operator at one station and his correspondent at the distant end. But not so if one arm is running faster or slower than the other, for then branch 2, at one station, might be joined either with 1 or 3 at the other.

Only an intermittent current, however, is sent over the circuit of each pair of operators; nevertheless, the pulses succeed each other with such rapidity that a practically continuous magnetic effect will be produced upon the relay in making a signal, provided the time required for an electro-magnet to part with its magnetism, upon the cessation of current, be longer than the interval between pulses.

The multiple-synchronous system, from a historical standpoint, is worthy of notice, not because of demonstrated superiority over other methods, but rather from the fact that it was the first multiple system invented. Moreover, it is important because of its promise of a capacity for a larger number of transmissions than it was supposed could otherwise be obtained. The public is occasionally startled with an announcement that some one has invented a telegraph by which a wire may be utilized for twenty or perhaps forty transmissions; but usually it is the old wanderer in a new garb. Speed by this method, however, is limited far within the bounds of these statements. It might seem that it would only be necessary to multiply the number of contacts and to increase the velocity of the rotating arms; but the limit in this direction is soon reached, for only a certain number of impulses can be transmitted over a line within a certain period with force sufficient to

10

produce signals. Many valuable improvements have been made in recent years in this class of telegraphy, but, large as the art has grown, the great object of all has been to obtain more perfect synchronism— that is to say, to cause two mechanically independent arms to rotate at the same speed.

The reduction of the duplex to practical form, in 1872, marked the most important advance in the art of telegraphy since 1844. For not only did it practically double the capacity of a wire by utilizing it for two simultaneous transmissions—one in each direction—but its development led to a careful investigation and a full understanding of the phenomenon of static induction on telegraph-lines.

In 1853 Dr. Wilhelm Gintl, of Austria, invented a duplex system which, in the following year, was so far improved by Carl Frischen, of Hanover, that it lacked only one essential element—means to balance the effects of static induction upon the relays—to bring it to its present perfection. This important addition was supplied by Joseph B. Stearns, of Boston, Mass., in the early part of 1872, and by its application the duplex became a successful means of doubling the telegraphic capacity of the longest circuits. From that moment messages were simultaneously transmitted between New York and Chicago, and upon lines of even greater length. Yet before this improvement the duplex was of no greater utility than had been the systems which had preceded Morse.

It is said that if Morse had failed in 1844, some one would have succeeded within a few years. It, however, required eighteen years to supply one step, or, more properly, to discover one fault in the duplex. at a time when its value was as certain as the fact that two telegraph lines cost more than one.

The principal characteristic of the duplex is, that a signal which is sent to a distant station for reproduction shall produce no effect upon the home receiving-instrument. In transmitting a signal, Frischen split the outgoing current into equal parts, and used one-half on the main line to produce a signal at a distant station, and the remainder upon an artificial line, beginning and terminating in the same office, to prevent signals at the home station. But, that the division of the current between the main and artificial lines may be equal, the resistance to the electrical flow in one must be made equal to that of the other. The use, however, of great lengths of wire for the artificial line is avoided, by employing a German-silver conductor of such small calibre that only a foot of its length may have the resistance of a mile of tele-

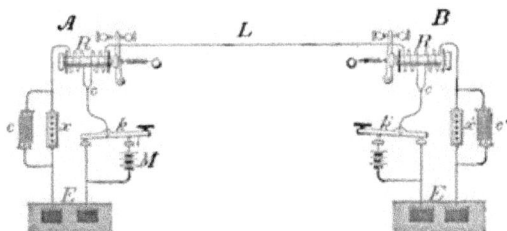

The Duplex System, for Simultaneously sending two messages, one in each direction, on a single wire.

In the accompanying diagram, showing the essential parts of Frischen's method as improved by Stearns, *A* and *B* are the respective stations at the opposite ends of the line. At *A* the artificial line is split from the main circuit *L* at a point *c*, and is made to include a second relay coil and an artificial resistance, *X*, and from the latter continues to the ground at *E*. Each coil has an equal number of turns around the relay core, but the one is wound in opposition to the other, so that currents of electricity passing simultaneously through them will create neutralizing effects. The resistance of the artificial line is made equal to that of the main line *L*, extending from point *c* of station *A* to earth, *E*, at station *B*, by adding resistance-coils *X*. If, now, key *K* be depressed upon its front stop, *I*, battery *M*, whose pole is connected with *E*, will be connected with the line, and the current issuing from the battery *M* will be divided at *c*, one-half flowing over the main line to produce a signal upon relay *R* at station *B*, while the other half passes through the artificial line, and thus acts upon the core of the differential relay *R*, to neutralize the magnetic effect produced by the main-line flow. A current in passing along a wire coiled around the iron core of a relay makes the core magnetic, and produces a signal through the attraction and consequent vibration of a movable iron bar, or armature, which is normally held in a back position by means of a spring.

graph-line; and by this expedient an artificial line which will balance a long telegraph-circuit may be reduced to the compass of a small box weighing only a few pounds.

If, with the batteries arranged as shown in the accompanying diagram, both keys were depressed at once, no current would flow over the main line, because one battery would oppose the other. Still, signals would be made at both stations, notwithstanding an absence of main-line current, for the relay cores would be made magnetic by currents in

the artificial line-coils. If, however, the battery at one end were
changed about, with its negative pole to line, its positive being connected with the earth, one battery would
not neutralize the other upon the depression
of both keys. On the contrary, the main
line current would be of double strength.
As before, signals would be simultaneously
made at both stations,
but from a different
cause. In this in-
stance the double cur-
rent in the main-line coils of the relays overbalances the
single strength flowing in each artificial line.

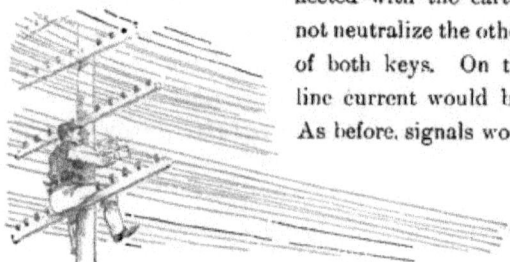

When a telegraph-line is connected with a battery in
sending a signal, it is charged, or filled from point to
point, as is the bed of a river with an unlimited supply
of water flowing from its source. A current would not
begin at any point until the bayous and lagoons above
had been partly or wholly filled, for they would serve as
reservoirs, temporarily, to exhaust the supply; and as
time is required for a flood to set in from the source to
the mouth, so there must intervene an appreciable period
before an electrical current of normal strength will be
established throughout the length of a telegraph-line.
While a line is becoming charged, a variable current,
starting at great strength at the battery end, is set up,
because the electric flow in the beginning encounters only
the resistance of a short length of line; and after the
removal of the battery, if the line is connected with earth, a momentary
return current will occur. Thus it is seen that, accompanying each
signal transmitted, an abnormally strong current will flow from the battery at the first instant, while at its termination there will be a strong
return discharge.

Linemen at
Work.

These currents were the source of difficulty in Frischen's duplex, for they were not balanced upon the differential relay, because there were no similar currents in the artificial line. A circuit made up of a short, thin wire, like Frischen's artificial line, however great its resistance to the electric flow, has no considerable metallic surface and no appreciable electrostatic capacity. Stearns gave the artificial line an electrostatic capacity, and thereby, at the beginning of a signal, the abnormally strong current flowing to the main line was balanced by an approximately equal one passing into the artificial line. Likewise at its termination the discharge from the main line was balanced by an equal return-current from the artificial line. And this Stearns accomplished by connecting the opposite plates of a Leyden jar, or condenser, one above and one below the resistance *X.*

In the diplex, forming a part of the quadruplex of the Western Union, two messages are simultaneously sent over a wire in the same direction, one by current-reversals and the other by changes in current-strength; and although both signals are electrically effected, they are, nevertheless, as independent as would be two messages, if one were sent along a metal rod by the blows of a hammer, and the other electrically by the Morse method. [See diagram and explanation on page 150.]

Doubtless many methods of illustrating double transmission will suggest themselves to the reader. A long rod might be moved backward and forward along its axis by one operator to ring a gong, while at the same time a second operator could rotate the rod about its axis to move a flag or to turn the hand of a dial. Two transmissions could also be effected by the action of water in a single pipe. If a section of the pipe were of glass, a valve placed within could be made visibly to move to and fro, and by the backward and forward flow thus caused, to indicate signals of one message, while signals of a second message could independently and simultaneously be indicated by increased pressure, shown by the height of fluid in a vertical pressure-gauge.

It has now been shown that two messages may simultaneously be transmitted in opposite directions, and it is shown below that two messages may simultaneously be sent in the same direction. It will readily

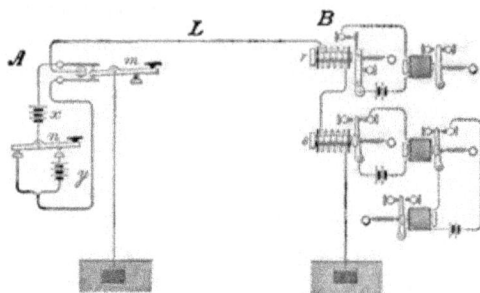

The Diplex System, for simultaneously sending two messages in the same direction on a single wire.

In the diagram of the diplex, two transmitting keys, *m* and *n*, are shown at station *A*, and two receiving-instruments, *r* and *s*, at station *B*. Receiving-instrument *r* is a polarized relay—or, for convenience of illustration, an electro-magnet having an armature consisting of a permanent magnet, responsive only to the backward and forward flow of current in the line, commonly known as reversals; while receiver *s* is neutral, its armature being of soft iron and therefore being actuated only by increasing the current strength. By the depression of key *m* the current normally flowing to line from battery *x* is reversed, and by the depression of *n* battery *y* is added to *x*; while, if both keys are simultaneously operated, not only is *y* added to *x*, but both are inverted and a strong reverse current is sent to line. A normal current of minor strength, sent to line when both keys are underpressed, magnetizes the polar relay in such manner as to force its armature to a back or non-signalling position. If, however, key *n* alone is depressed, a full current of normal direction will be transmitted which will serve to press the armature of the polar relay more firmly to its back position. If the direction of current normally flowing develops a north magnetism in the right end of the relay core, the upper or vibrating end of the armature being north, one mutually repels the other : if, however, the current is reversed, the resulting south magnetism of the iron core and the north magnetism of the armature become mutually attractive. An increased current, therefore, will not actuate the polar relay to produce a signal, but it will serve to magnetize the core of the neutral relay sufficiently to overcome the strong retracting spring of its armature, and thus to produce a signal. The neutral relay is equally influenced by positive and negative currents, the armature being of soft iron and equally attracted by either north or south magnetism, and hence it is that the reversal of a minor current which is capable of moving the polar relay will have no effect upon its armature. If, therefore, the latter be in its back position, it will not be actuated by the reversal of a minor current. Likewise, if the armature of the neutral relay be attracted, a subsequent reversal should not cause its movement. In practice, however, it is found upon long lines that during reversal there is a tendency for the armature of the neutral relay, if attracted, to be drawn back by its spring and thus to mutilate its signals. In other words, the period of static charge and discharge during a reversal is so long that the neutral relay, which should be wholly governed by changes in current-strength, has the fault of responding to reversals and to signals sent by the wrong operator. This is the great difficulty encountered in operating the quadruplex system. Otherwise it would be as perfect in its operation as the single Morse system.

be understood [Diagram, page 151] that the quadruplex, by which two messages are simultaneously sent in each direction, is formed by placing at each end of the duplex two transmitters and two receiv-

ers such as are shown in the diplex. In this arrangement the artificial
line of the duplex is made to include a neutralizing coil on each of
the two relays, thus prevent-
ing the receiving-instruments
at the home station from responding to outgoing signals,
while the reversing and cur-
rent-changing keys independently serve to bring into action the polar and neutral
relays at the distant station.

Instead of transmitting
two messages in the same
direction, one by reversals
and the other by changes
in current-strength, both
transmissions may be effected by employing three different strengths of current,
all in one direction: and,
in fact, this principle has

Diagram of the Quadruplex Telegraph, for sending four messages, two in each direction, at the same time on a single
wire.

been adopted in using the quadruplex as a foundation for a sextuplex
—a system for three simultaneous transmissions in each direction.

In the multiple-harmonic system, by which many messages may be
transmitted in one direction, or in opposite directions, each operator,
by depressing a Morse key, puts in action a vibratory circuit-breaker,
and thus causes a series of electrical pulses to flow over the main
line and through several receiving electro-magnets, which are provided
with vibrating armatures formed of reeds or steel ribbons so propor-
tioned that their different rates of vibration may be made equal to those
of the transmitters. If the elastic reed or ribbon of a receiving-instrument is not tuned to vibrate in unison with the transmitter it will not
be brought into action, but will remain quiescent, as would a pendulum

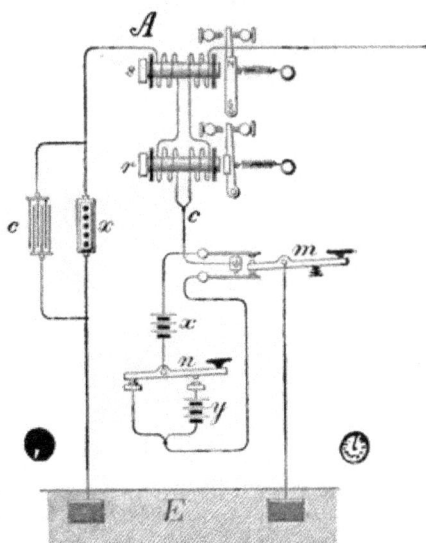

if forces were applied on both sides without regard to its period of vibration and direction of movement. The several receivers are thus made responsive to the corresponding transmitters, while each is silent to all but its own pulsations; and although a composite tone will be transmitted when all the sending keys are simultaneously depressed, no interference between the several transmissions will ensue; for each receiver, under the action of the resultant series of pulses, is vibrated,

S. F. B. Morse.

as though only an intermittent current from its own transmitter were sent to line. The several receivers act to analyze the composite series of pulses, each taking up a component series equal in number to the vibrations derived from its transmitter. If the several transmitters were tuned to the notes of a musical scale, a tune could be played and reproduced by the several receivers if placed in the same room; but each receiver would produce only its characteristic note, as is found by placing them in separate apartments. Thus an independent message

may be transmitted by each of the several keys, and it will be reproduced only upon the corresponding receiver.

The efficiency of this system, however, is seriously impaired by
inductive disturbances from other wires on the same poles, and probably this defect, more than any other, has prevented its adoption.
Experiments, however, at moderate distances, with only one wire on a
line of poles, seem to have been very successful.

Morse originally proposed to employ type-blocks, which, placed in
forms, were mechanically moved under the arm of a circuit-breaker, to
automatically transmit signals. In 1832 he suggested also the electro-
chemical process of discoloring a strip of paper for making a permanent
record. But neither idea was practically applied by him. The modern
chemical automatic was first put into experimental form by Bain before
1850, but with little success. Bain, as is now done, transmitted messages by drawing a perforated strip of paper between the points of a
key and a metallic surface, the holes in the paper permitting the two to
come in contact, and thus to transmit a signal. Morse did not suggest
the automatic for the sake of great speed—he only sought mathematical
accuracy in transmission. Whatever Bain hoped to accomplish in the
direction of greater capacity, his primary object was to make a telegraph that he could use notwithstanding Morse's patents. In 1869,
however, the electro-chemical automatic was brought to public notice as
a system possessing rare qualities of speed. But in its several competi-
tive trials with the Morse it has proved a remarkable failure, although,
perhaps, more than a moderate degree of success might have been
expected. Many believed that it would give a wire at least thirty times
the capacity of a Morse circuit; and perhaps not without reason, for
President Grant's annual message of 1876 was sent over the wires of
the Atlantic & Pacific Telegraph Company, from Washington to New
York, at a rate which apparently justified this estimate. But, notwithstanding the theoretical advantages of the system, it has failed in the
hands of companies having the strongest financial support, and, in fact,
it has ruined every organization which has persisted in using it in com-

petition with the Morse.   And so conspicuous have been the failures
that their history may have some interest for the general reader.

    We may obtain an idea of the enthusiasm aroused in behalf of this

Sending Coffee Quotations over Ticker Circuit.

system from the annual report of Postmaster-General Cresswell, of
November 14, 1873 :

    " For years past the attention of inventors and scientists has been attracted to the
necessity for a more rapid and less expensive mode of transmission than the Morse,

which requires the messages to be spelled out by a slow and tedious process, at **about** the speed of an ordinary writer. One of the results of their investigations is the automatic or fast system now in operation between New York and Washington. This system is capable of a speed of from five hundred to eight hundred words per minute. The

```
 LI          ET N            UT.EX         E.II               .
  4S.2.99⅝      5S.2.105¼           103₄B       103⅜.....⅝
```

Stock Quotations as Received on the Scott Instrument.

```
 .DL...           .DL....           .NPR....             ,
 ,....200.140⅛S3.....,700.140⅛, ,...200.50⅔.
```

Specimen **from the** Edison **Stock** Printer.

average of an expert **Morse operator** is not over twenty-five words per minute. Therefore it is evident that if the automatic method can be made **to** accomplish what its advocates **confidently** predict for it, the capacity of a single wire for business will be increased **nearly or quite thirty times.** . . . There can be no **doubt of** the ultimate success of the automatic principle. Its battle with an incredulous public is almost **won.** As soon as it shall be thoroughly developed and applied in **practice the problem of** cheap telegraphy will be definitively solved."

The Postmaster-General assumed that the time had come for the formation by the Government of a postal telegraph system. Failing, however, to induce Congress to build lines, the owners of the automatic system, in 1874, secured a purchaser in the Atlantic & Pacific, a corporation owning many thousand miles of wire, which was then in opposition to the Western Union. Notwithstanding the fact that the Western Union had recently doubled the capacity of its wires by using the duplex, it was naturally assumed that, with the assistance of the chemical automatic, the Atlantic & Pacific could operate its lines at a profit after reducing Western Union rates by one-half, for it had been confidently represented that a wire thus equipped was capable of at least thirty Morse transmissions. President Grant's message in 1876 had been telegraphed two hundred and fifty miles at the rate of several hundred words a minute; but, notwithstanding this and other apparently successful tests, the system, after a use of about two years, was discarded. The failure was made conspicuous by the fact that the Atlantic & Pacific was able to give the undertaking all necessary support. This, therefore, is not an instance of a meritorious invention permitted to perish for want of nourishment in infancy. It was not abandoned until it had proved an expensive experiment; and in the end its worthlessness was so thoroughly demonstrated that the Atlantic & Pacific, having only an imperfect system of double transmission, was for the most part reduced to the use of single Morse instruments, while its rival enjoyed the advantage of the Stearns Duplex.

Again, in 1879, an automatic system containing many valuable improvements was taken up by the American Rapid Telegraph Com-

I AM A BULL ON THE STOCK MKT AND WOULD CERTAINLY BUY SHARES IN

Kiernan's News Tape.

pany, a corporation of large means, whose lines were built to remedy certain defects said to have contributed largely to the Atlantic & Pacific failure. But after a trial of nearly five years the automatic was again abandoned. The American Rapid began with a system which, of its class, will probably never be excelled, and for which it was promised that two thousand words a minute, instead of one thousand, could be sent over a wire; but at one speed or another, with good wires or bad, the automatic system seemed equally potent to break down any company attempting to use it to the exclusion of other methods.

In 1883 an effort was made by the Postal Telegraph Company to introduce the automatic system of Leggo, upon a large wire of low resistance, between New York and Chicago; and although, for the purposes of an automatic, this was probably the best line ever built, the results were so unsatisfactory that after experimental use for about three years the system was finally abandoned.

It is now maintained by advocates of this method of telegraphy that four thousand words a minute may be sent over a single wire; but, considering the signal failures at one thousand and two thousand, these assertions only lead to the conclusion that the great speed of the system is of no avail, and that it is the *ignis fatuus* of the telegraph world.

New men will from time to time be induced to take up this chimera

Perforating Messages to be Sent by the Wheatstone System.

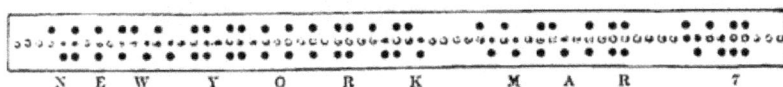

Spec-men of Perforated Slip for Sending Message by the Wheatstone; same message as received.

as a means of revolutionizing telegraphy, but a company could now wish a competitor no greater harm than the use of an electro-chemical system as its principal method of transmission. The automatic is doubtless a valuable auxiliary to a telegraph system, but it cannot be exclusively used to advantage.

The Wheatstone telegraph is a system which has long been used with a high degree of success in Great Britain, and has in late years proved a valuable adjunct to the Morse in the Western Union service, particularly where large volumes of business must pass over few wires. In this system messages are automatically transmitted by a strip of perforated paper, while their reception is effected by an ink-marker which, under the action of a receiving electro-magnet, makes Morse dots and dashes upon a moving band of paper. Although the use of an electro-magnetic receiver makes impossible the high speed which may be obtained by the electro-chemical method, the one possesses advantages over the other which are indispensable to a successful system. In the Wheatstone, repeaters which serve to convey transmissions from one circuit to another without manual aid may be employed, as is done at four points on a line twenty-six hundred miles long, from Chicago to San Francisco, while in the electro-chemical system this is impossible; and for this reason alone it is not practicable upon lines of the greatest length, where it would be most useful. Moreover, the record, when made in ink-marks, is far more reliable than when formed by electro-chemical discolorations on moistened paper, for in the latter case, at great speed, the tendency for dots and dashes to become blended into a continuous line is marked. The Western Union has long controlled the electro-chemical systems with which the Atlantic & Pacific and the American Rapid Companies failed; but it has not attempted to utilize either, and most of the apparatus has long since found its way to the junk-dealer.

The *fac-simile* telegraph, by which manuscript, maps, or pictures may be transmitted, is a species of the automatic method already described, in which the receiver is actuated synchronously with its transmitter. By Lenoir's method a picture or map is outlined with insulating ink upon the cylindrical surface of a rotating drum, which revolves under a point having a slow movement along the axis of the cylinder, and thus the conducting point goes over the cylindrical surface in a spiral path. The electrical circuit will be broken by every ink-mark on the cylinder which is in this path, and thereby corresponding marks are made in a spiral line by an ink-marker upon a drum at the receiving end. To produce these outlines it is only necessary that the two drums be rotated in unison. This system is of little utility, there being no apparent demand for fac-simile transmission, particularly at so great an expense of speed, for it will be seen that instead of making a character of the alphabet by a very few separate pulses, as is done by Morse, the number must be greatly increased. Many dots become necessary to show the outlines of the more complex characters.

The pantelegraph is an interesting type of the fac-simile method. In this form the movements of a pen in the writer's hand produce corresponding movements of a pen at the distant station, and thereby a fac-simile record.

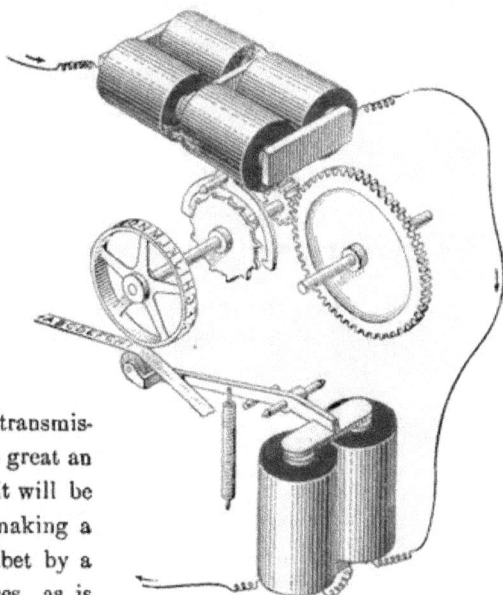

Type-printing Telegraph for Distributing Quotations and News on Short Lines.

The diagram indicates the principal parts of a step-by-step printer which, in various modifications, has been very generally used for reporting quotations and news upon short lines in cities. The type-wheel is rotated in this case by a clock-motor, and its step-by-step action is limited by reverse currents sent over the circuit. If a short pulse, of one polarity or the other, is prolonged, a neutral magnet in the same circuit is actuated to press a paper strip against the wheel to effect printing. To print a particular character, therefore, it is only necessary to transmit a number of reversals; to turn the wheel from the position which it last occupied, so bringing the character over the press-pad, then to prolong the last current transmitted, to effect an impression.

In the many forms of type-printing telegraphs which have come into use there has been employed a rotating type-wheel carrying the necessary characters, as shown in the accompanying illustration.

Printers in which the type-wheel is rotated step by step were, in the earlier days of telegraphy, employed upon comparatively long lines, and

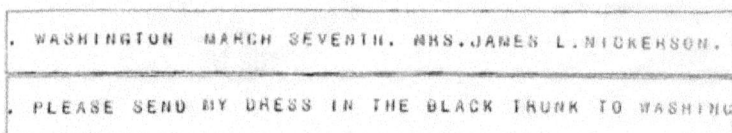

```
. WASHINGTON    MARCH SEVENTH.  MRS.JAMES L.NICKERSON.
```

```
. PLEASE  SEND  MY  DRESS  IN  THE  BLACK  TRUNK  TO  WASHING
```

Telegram as Received by the Phelps Motor Printer.

very considerable speed was obtained, but because of the number of pulses required to bring the type-wheel into position for an impression such printers are not well adapted to the longest lines, the speed at which one pulse may be made to follow another being limited. [Page 159.]

In the Phelps motor printer, which is used to a limited extent by the Western Union Telegraph Company, only one pulse transmitted over the main line is required to print each letter. This is accomplished by the synchronous principle, a transmitter at one station and a type-wheel at the other being rotated by suitable motors at exactly the same speed. In the rotation of the transmitter, upon depressing a key, a current is sent to line just when the character on the type wheel corresponding with the key depressed is brought opposite the press-pad.

Only the feeblest currents should be used on submarine lines, since heavy pulses which could be employed with impunity on land lines, if they did not soon destroy the cable-covering, would at least tend to develop faults which otherwise might long remain latent. Defects in cable-covering, that otherwise may not lead to harm, admit moisture, and hence, under the action of a strong current, oxides are quickly formed, destroying insulation. The necessary use, in ocean telegraphy, of the lightest currents has led to the development of a class of recording instruments remarkable for delicacy of action—notably the Siphon

Recorder, which indicates the electric impulses by a wavy ink-line on a tape, and the Reflecting Galvanometer, which causes a spot of light to move from right to left in a darkened room. With these recorders and thirty cells of battery, messages sent across the Atlantic are telegraphically reproduced in ink at the rate of from twenty to twenty-five words a minute, each way, the cable being duplexed. But for electrostatic induction a single cell of

The Siphon Recorder for Receiving Cable Messages—Office of the Commercial Cable Company, New York.

battery would suffice for transmission from the earth to the moon, if those bodies could be connected by a wire of the size used in ocean cables. Indeed, under such conditions, if there were only the resistance of the wire, with the larger batteries now used in working the quadruplex on land lines five or six hundred miles long, messages might be sent from the earth to the sun, or from one planet of the solar system to another. The impediment of static induction in telegraphy is strikingly exemplified in the ocean

11

telegraph. If there were no such phenomenon a single battery cell could operate around the globe better than do thirty cells across the Atlantic. Upon land lines there is usually found not more than one-fiftieth of the opposition encountered from this cause on ocean cables; yet, even here, the amount of current absorbed along the line by static induction is far greater than the portion employed in making signals.

While Morse in planning his telegraph apprehended that the fluid might not act upon a circuit with sufficient force to effect transmission over great distances, he was confident of his ability to accomplish this, if a line could be worked ten miles, by employing a

Locating a break in an Atlantic Cable—Office of the Commercial Cable Company.

series of circuits joined by relays. His assumption of only one prime difficulty, which was, in fact, imaginary, is instructive, when we consider how small a battery is now required to work a cable recorder at distances exceeding his most sanguine speculation.

Simple as are the methods of locating mid-ocean breaks in cables, so that a vessel may sail to the point of rupture, they are, perhaps, not popularly understood. If the metallic conductor were broken, the surrounding insulation remaining perfect, the electrostatic charge of the cable, or the amount of electricity which it absorbs in becoming charged, is electrically weighed by building up an artificial line until the current flows equally into the cable and such artificial line. This equal division of the current might be indicated by a differential instrument resembling the duplex relays, shown on page 147. If the currents are equal, the armature of the differential instrument remains quiescent during their passage through its coils, for one balances the other. Each element of the artificial line having the static capacity of a known length of cable, an inventory of the elements used in making the artificial line would obviously give the length of the cable. This process is no more abstruse than would be the determination of the distance to a water-tight obstruction in a pipe, by forcing in or pumping out water, and taking its measurement. If it were known that a pint of water

Section of Cable, Chafed and Torn by an Anchor.

filled a foot of the pipe, the length of the conduit would equal in feet the number of pints pumped out or forced in.

If, however, the insulating covering of the wire is broken, the current will flow freely from the conductor to the surrounding water, and its strength, if the power of the battery is known, definitely measures the electrical resistance, and consequently the length of the conductor. [See Ohm's law in "Electricity in the Service of Man."] In other words, the battery-power, divided by the indicated current-strength, gives the line's resistance, and therefore its length. In the same manner, if we know the head of a water-supply, we may easily determine the length of a pipe by noting the velocity of the flow.

Few discoveries have added more to the fund of electrical science than the recent determination of the fact that feeble telegraphic currents may inductively be conveyed across an air-space of one hundred feet or more.

The idea of telegraphing to moving trains had its inception as early as 1858; but of the many forms suggested all were impracticable in that they involved a mechanical contact between the train and the stationary conductor. Obviously, it is not feasible to make a circuit, either through a sliding arm projecting from a car, or by so modifying the track of a railroad that its rails may be utilized as electric conductors. But that this may be done by induction there can be no doubt, for its feasibility has been shown in daily practice upon the lines of the Lehigh Valley Railroad for the past three years. A moving train may now receive messages passing along a neighboring wire almost as readily as New York communicates with Philadelphia by ordinary methods.

Barnacles on a Cable.

Nor does the great speed of the train interfere with successful communication. If it could attain the velocity of a meteor, signals upon the wire would fly across the intervening space, inductively impressing themselves upon the metal roofs of the cars, with the same certainty as if the cars were motionless upon a side track; and it is not even essential that the train and the line be separated by a clear air space, for non-conducting or non-magnetic substances may be interposed without impeding transmission. During the memorable blizzard of March, 1888, the capacity of the system, in this particular, was subjected to an instructive test on the Lehigh Road.

On the afternoon of March 14th, the second day of the storm, an effort was made to clear the road by forcing a train of four locomotives and two tool cars, carrying two hundred workmen, through a long cut near Three Bridges Station. The snow had filled the cut upon the north side high above the train and the telegraph wire, but the south side was nearly empty. As a consequence of the great resistance offered by the snow on one side, the three locomotives in the lead were forced off the rails to the right, and were badly wrecked. No sooner had the train operator recovered from the shock, than he sent a message advising the division superintendent of the accident, and requesting aid for the killed and injured. For a long distance, at this point, snow and ice covered the wire to a depth of more than ten feet, yet during the following three days, in clearing the wreck, about six hundred messages were sent between the car on the south track and the wire running along the north side of the road. Notwithstanding the intervention of snow, communication remained as clear as before.

When an electric current passes over a telegraph-line, objects along its length, although at a considerable distance, are electrostatically charged, and thereby secondary currents are made to flow between these objects and the earth, both at the beginning and at the end of the electrical signal. Thus, if at the beginning of a signalling pulse a secondary current flows over an electrical conductor from a car-roof to the earth, at its termination a return-discharge will be established from the earth to the roof. But the currents between the roof and the earth are only momentary, and they will not be lengthened by prolonging the flow upon the main line. For this reason it is necessary to send many pulses over the main line, in making either a dot or a dash of the Morse alphabet, in order that a continuous effect may be produced upon the instruments in the car. If fifty short currents, following each other in rapid succession, should pass over the main line during the depression of a key to make a Morse dot, or one hundred to make a dash, double the number of induced pulses in the train-circuit would be developed, and tones like the buzzing of a wasp, of varying duration, representing dots and dashes, would follow in the receiving-telephone, which in this case replaces the Morse sounder. These signals are taken up by the cars all along the line, whether upon one track or the other of the road, and when running in either direction.

The accompanying diagram illustrates the arrangement of the apparatus used in a car. As here shown, an intermittent current is inductively thrown upon the line during the depression of key *K*. When the key is closed, primary circuit 6, 7, 8, 9 of an induction coil, including battery *B*, is first closed, and thereupon is rapidly opened and closed by an automatic vibrator, consisting in part of an armature, *A*, which normally rests against back contact, *n*. The armature is first attracted, breaking the circuit; the iron core of the induction coil is then demagnetized, permitting the armature to return to contact *n*, and again to close

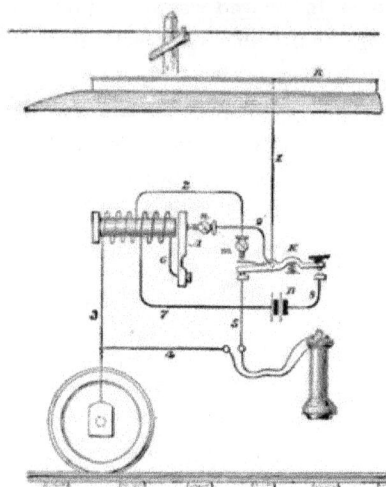

Diagram showing the Method of Telegraphing from a Moving Train by Induction.

the primary circuit. In this manner the forward and backward movement of armature *A* is effected. The secondary circuit of the induction coil, which is wound upon the same core and insulated from the primary, is shown by wires 1, 2, 3, and, although normally open, is closed upon the depression of the key at contact *m*, and is made to connect the car-roof *R* with the earth through wheel *W* and rail *M*. An induced current is set up within the secondary coil at each opening and each closing of the automatic vibrator, and as a result the car-roof is electrified, first above and then below the potential of the earth with which it is connected, and in each case its condition is impressed by induction upon the wire at the roadside. In this manner

intermittent currents are thrown upon the line and are received upon a
telephone at a distant station by "buzzing sounds," as in the car.
When signals are not being sent from the car the transmitting key is
not depressed, or is in its back position, and a receiving telephone in
the car is placed in circuit between the roof and the earth. At stations
on the line the transmitting apparatus is not unlike that already described, except that the secondary circuit of the induction coil is made
a part of the main line, while the telephone receiver is placed in
the primary circuit of the induction coil.

The train system of the Lehigh Valley Railroad is provided with a

Train Telegraph—the message transmitted by induction from the moving train to the single wire.

single wire suspended upon a line of poles about sixteen feet above the
rails and some ten feet from the double roadway, and hence about eight
feet from the cars on one track, and twenty feet from those of the other;
and so perfectly does the system act, at varying distances, that communication is quite possible even to a fourth track from the wire. While
any one of the large collection of wires usually built along a railroad
may be used, it is preferable, as is done by the Lehigh Valley Company, to employ a wire isolated from the others and carried by a separate pole line. The line, however, need not be wholly given over to
train telegraphing, for messages may simultaneously be transmitted by
the Morse system, thus duplexing the wire.

The outfit of an office upon a train consists essentially of a telegraph-key, with an induction coil mounted upon a small board, which the operator holds in his lap; a telephone carried at his ear, and a hand-case containing a few small cells of battery. To equip a car, therefore, is only the work of a moment. One wire is passed through the bell-cord

Interior of a Car on the Lehigh Valley Railroad, showing the Method of Operating the Train Telegraph.

opening above the door and is attached by a clip to the end of the metal roof, while a second is passed out of the window and fixed to a metal part of the car-truck. The instruments are then properly arranged in circuit between the roof and the earth, and the operator is prepared to send or receive messages. As all passenger cars have long been roofed with tin, the only necessary outlay is a few cents for clips wherewith to attach the wire to the roof and to the truck.

In 1881 William Wiley Smith, of Indiana, proposed to communicate between moving cars and a stationary wire by induction; but he appears to have thought this practicable only at small distances. The remarkable possibilities of long-distance transmission by induction seem to have been discovered by T. A. Edison and L. J. Phelps, who, aided by others, have added improvements adapting it to every requirement of a commercially successful telegraph system.

The train-telegraph of the Consolidated Railway Telegraph Company, which has just been described, is now used upon more than two hundred miles of the Lehigh Valley Railroad, and although it has not found wider use, it deserves, as a scientific achievement, a high place among electrical improvements of the last ten years. It certainly would seem to be of great value in operating trains, for it offers a means whereby a despatcher may instantly communicate with every train under his supervision.

The inductive method of transmission employed in train-telegraphy has already been suggested as one way of telegraphing over considerable distances without the aid of a wire. It is said that Edison has succeeded in obtaining intelligible signals between apparatus placed upon masts five hundred and forty feet apart, and it is now confidently asserted that messages may be telegraphed from a wire along the Hudson River to steamers plying between New York and Albany. It is claimed by Mr. L. J. Phelps that by employing a receiving-coil consisting of a wire wound several times around the steamer, one of the early forms employed in the train-telegraph, no difficulty would be experienced in communicating with boats running within five hundred feet of the shore; and that such communication would be possible even over a distance of two thousand feet. Obviously, by this method, ships not far apart may be placed in communication.

Telegraphing between points not connected by a line is also accomplished by diffusion. If an electrical current is sent over a telegraph-line connected with the earth at both extremities, it will return through the earth from one extremity to the other. In returning, the current

will not follow a narrow, straight line, but will be diffused.; and although
the larger part will return by the easiest and shortest path, enough to
actuate a telegraphic apparatus may flow back by way of a distant
parallel wire. Two wires, one along the shore of the Isle of Wight, and
the other along the shore of the channel, in Hampshire, were thus

Check-girls who Collect and Distribute Messages—Western Union Main Operating Room, New York.
[Before the Fire.]

employed ; and although the width of the channel is six miles at one
end, and one and one-fourth mile at the other, harmonic signals like
those used in the train-telegraph, sent on one wire, were distinctly
reproduced upon the other, telegraphic communication across the channel being thereby established.   Professor A. G. Bell succeeded in tele-

graphing in this manner more than a mile between boats on the Potomac River.

Respecting the action of diffusion, it should be noted that the battery in a telegraph-line, when the latter is closed, tends at one end of the line to raise the electrical potential of the earth, and at the other to lower it. If an engine in a long tunnel were pumping air through the tunnel, bringing it in at one end, and forcing it out at the other, a barometer would show higher pressure at the point of egress than at the point of entrance, and, as a consequence, currents of air would tend to flow back to the entrance through the outer atmosphere, covering in their return a wide tract of country. If such an operation should be carried on in a tunnel extending from New York to Chicago, it is not to be presumed that windmills in Kentucky would be much disturbed, although in theory such a tendency would exist.

Had the promoters of the telegraph foreseen the extent of its subsequent improvement, they would then have considered its perfection as substantially assured. But the modern telegraph doubtless seems to us even further from perfection than did Morse's system appear to him when he had first succeeded in working from Washington to Baltimore. For the telegraph might now have almost unlimited speed if difficulties which have become apparent in the course of its development could be successfully removed.

The capacity of an Atlantic cable has been advanced from ten to fifty words a minute; but valuable as are the immediate results of this achievement, they have served the better purpose of making clear that if the impediment of static induction could be eliminated, transmission of five hundred words a minute would be equally possible. Stearns's discovery doubled the capacity of every long line in the world, and made possible the quadruplex. Yet his improvement was more valuable in other ways; for when it had been widely adopted the phenomenon of static induction was brought under the daily observation and scrutiny of hundreds of intelligent men in a manner to insure its removal, if possible. From Stearns's invention it became a matter of common information that all along a line, with each signal transmitted, the electrical conditions of a thunder-cloud are feebly reproduced. For illustration, we may assume that from the wire hangs an invisible electrical cloud from which, if the tension of its charge were sufficiently increased, lightning flashes would burst forth through the air to the earth.

Perhaps static retardation and the absorption of current in electrifying the surrounding air along the line are as inseparably connected with telegraphic transmission as is friction with machinery, and a solution of the problem may be an impossibility. But the expedients of raising the line on poles high above the earth, and of dividing a long circuit by repeaters, show at least two methods by which the difficulty may be modified, if it cannot be overcome, and encourage us to hope that better means may be found. It is certain that fame and fortune await him who shall solve the mystery.

THE MAKING AND LAYING OF A CABLE.

By HERBERT LAWS WEBB.

THE WORLD'S SUBMARINE TELEGRAPH SYSTEM—TWO HUNDRED MILLIONS OF CAPITAL INVESTED—THE GREAT CABLE FLEET—UNIQUE LIFE ON A CABLE SHIP—A TYPICAL VOYAGE—HOW THE CABLE IS MADE—THE TEST FOR LEAKAGE—SHIPMENT OF THE CABLE—IN THE TANK—THE PREPARATIONS CONCLUDED—SAILING AWAY—AN INTERESTING BODY OF MEN—ENGINEERS AT WORK—LANDING THE SHORE END— A CABLE HUT—PAYING OUT—IN THE TESTING-ROOM—SCENE ON DECK—BUOYING THE END AT SEA—ON THE WAY TO THE CANARIES—AN OCEAN SURVEY—THE SOUNDING MACHINE—DISCOVERY OF A SUBMARINE MOUNTAIN—RECEPTION AT LAS PALMAS—A SPLICE—A DOUBLE DISASTER—GRAPPLING FOR A LOST END—THE FINISHED WORK.

IN these days of rapid development in new fields of electrical science and their commercial application, it is easy to overlook the magnitude of the work accomplished in the laying of deep-sea cables. According to the latest report of the International Bureau of Telegraph Administrations, the submarine telegraph system of the world consists of 120,070 nautical miles of cable. Government administrations own 12,524 miles, while 107,546 are the property of private companies. The total cost of these cables is in the neighborhood of two hundred million dollars. The largest owner of submarine cables is the Eastern Telegraph Company, whose system covers the ground from England to India, and comprises 21,860 miles of cable. The Eastern Extension, which exploits the far East, has 12,958 miles more. Early in last year the system of West African cables, which started from Cadiz only six years ago, was completed to Cape Town, so that the dark continent is now completely encircled by submarine telegraph, touching at numerous points along the coast.   More than 17,000 miles of cable have been required to do this, and several companies, with more or less aid from the British, French, Spanish, and Portuguese governments, have participated in carrying out the work.

The North Atlantic is crossed by no less than eleven cables, all laid since 1870, though I think not all are working at the present time; five companies are engaged in forwarding telegrams between North America and Europe, and the total length of the cables owned by them, including coast connections, is over 30,000 nautical miles.

The cable fleet of the world numbers thirty-seven vessels, of an aggregate gross tonnage of about 54,600 tons.   Ten ships belong to the construction companies, their aggregate gross tonnage being about half that of the entire fleet; the other twenty-seven are repairing-steamers belonging to the different government and telegraph companies; they are stationed in ports all over the world, keeping a watchful eye on the condition of its submarine nerves, and doctoring them up whenever they need attention.   The *Silvertown* and the *Faraday* head the list of cable ships in point of size, the former being 4,935 tons, and the latter 4,916 tons; while the *Scotia* (an old Cunarder) is a close third with 4,667.   The *Faraday* has laid several of the Atlantic cables, and the *Silvertown* has done a great deal of work on both coasts of South America and on the west coast of Africa.   This ship has exceptional capacity for carrying cable, her main tank being fifty-three feet in diameter and thirty feet deep, large enough to stow a good-sized house in.   On one expedition she carried 2,370 knots of cable, weighing 4,881 tons, the whole length being coiled on board in 22 days, or at the rate of over 100 knots a day.   Better still, she laid the whole length without a single hitch, much of it being paid out at the high speed of nine knots an hour.

Among the repairing ships the best known is the *Minia*, the Anglo-American Telegraph Company's steamer, which patrols the North

Atlantic, keeping the many ramifications of submarine cables which radiate from the Newfoundland and Canadian coasts in working order.

The life on one of these cable vessels is unique and most interesting, combining the adventures of voyaging with operations demanding the highest scientific skill and knowledge, and with the most ingenious mechanical work. The men brought together are, of course, of widely varied experience and accomplishments, each in his way an expert in some branch of electrical or mechanical engineering. It was the writer's good fortune, in 1883, to be connected with the technical staff of such a vessel—the cable-ship *Dalmatia*—and he hopes that this narrative of his experiences will give a pleasant insight into the work of constructing the costliest and most wonderful half of Puck's girdle round the world.

In the summer of that year the Spanish Government decided to establish telegraphic communication between the group of Atlantic islands known as the Canary Islands, and the Spanish Peninsula, by means of a submarine cable, and also to connect various of the principal islands of the group with each other by the same method. This important work was entrusted to a leading English cable manufacturing company with a very long name, commonly called for short, "The Argentville Company," from the name of the place where the company's works are situated. It was for the purpose of laying these cables that the *Dalmatia* and *Cosmopolitan* made the voyage which I shall describe.

Let us first see what a submarine cable is, and how it is made. To do this a visit must be made to the enormous factory on the banks of the Thames, a few miles below London. Here the birth of the cable may be traced through shop after shop, machine after machine. The foundation of all is the conductor, a strand of seven fine copper wires. This slender copper cord is first hauled through a mass of sticky, black compound, which causes the thin coating of gutta-percha applied by the next machine to adhere to it perfectly, and prevents the retention of any bubbles of air in the interstices between the strands, or between the conductor and the gutta-percha envelope. One envelope is not sufficient, however, but the full thickness of insulating material has to be attained by four more alternate coatings of sticky compound and plastic gutta-percha. The conductor is now insulated, and has developed into " core." Before going any further the core is coiled into tanks filled with water, and tested in order to ascertain whether it is electrically perfect, *i.e.*, that there is no undue leakage of electricity through the gutta-percha insulating envelope.

These tests are made from the testing-room, replete with beautiful and elaborate apparatus,* by which measurements finer and more accurate than those even of the most delicate chemical balance may be made. Every foot of core is tested with these instruments, both before and after being made up into cable, and careful records are preserved of the results.

After the core has been all tested and passed, the manufacture of the cable goes on. The core travels through another set of machines, which first wrap it with a thick serving of tarred jute, and then with a compact armoring of iron or steel wires, of varying thickness according to the depth of water in which the cable is intended to be laid. Above the armoring, in order to preserve the iron from rust as long as possible, is applied a covering of stout canvas tape thoroughly impregnated with a pitch-like compound, and sometimes the iron wires composing the armor are separately covered with Russian hemp as an additional preservative against corrosion.

The completed cable is coiled into large circular store-tanks, where it is kept for some time submerged in water and again subjected to an exhaustive series of electrical tests. These tests form, so to speak, the baptismal record of the cable; by them it is ascertained whether the specifications have been complied with in respect to the maximum conductor resistance and the minimum insulation resistance which the cable

* A set of testing instruments for submarine cable-work, somewhat less elaborate than used in a cable factory, is illustrated on page 162.

is to have; in other words, whether the limits set by the purchasers of the cable on the amount of resistance in the conductor to the flow of the current, and the amount of leakage through the insulating envelope, have been exceeded or not.

The shipment of the cable next claims attention. The cable-steamer is lying at her moorings some distance out in the river, taking in her priceless cargo; and it is safe to say that the loading of no other ship presents such a curious and interesting scene. The cable is undulating in the air like an enormous eel as it emerges from the factory on the river-bank and travels over guides mounted on tall floating frames until it reaches the ship's side, over which it glides and immedltaely dives down into the dark recesses of the hold, where a gang of men are busy coiling it away, at the rate of four or five miles an hour, into one of the four iron tanks with which the ship is provided.

On board the ship there is a scene of confusion. The deck is strewn with packing-cases galore; stores of every description, some for use on board, others comprising complete equipments—from heavy furniture down to buckets and brooms—for the telegraph stations which the cable is presently to call into existence, coils of wire, huge spools or drums of underground cable (similar to those which have lately become familiar objects on the streets to every New-Yorker), to connect the landing-places of the submarine line with the town offices, galvanized iron cable-huts to be erected for the reception of shore-ends and instruments at these landing-places, tools of every description, huge iron buoys, coils of rope and heavy chain, grappling-irons and mushroom-anchors, cases of instruments, and formidable-looking trays of electric batteries; all these myriad objects—many of them labelled with queer-sounding Spanish names indicating their ultimate destination—surround one on all sides, as the work goes on of taking them on board and stowing them away in their proper places, there to remain until the hour arrives when they shall be called into action or unloaded in distant ports, to undergo stern and critical examination at the hands of

12

grave and dignified, or perhaps fussy and exacting, Iberian custom-house officials.

The cable—which, after all, is the principal character in this varied scene—is being dragged on board by steam machinery in a sluggish, hesitating sort of manner., Perhaps it is being coiled away into one of the tanks somewhat distant from the engine which is hauling it on board; in which case it is guided to the hatchway above the tank by means of grooved pulleys and long wooden troughs provided with little iron rollers, over which it rattles and whirs merrily.

In order to see the most important passenger that the ship is to carry installed in the depths of the dark, capacious state-room provided for its accommodation, it is necessary to take a peep between decks, and find one's way to "tank square," as the square opening on the main deck above the tank is called. Arrived at the tank in action, and standing at its edge, one can peer down into the gloomy depths; overhead a large grooved wheel, fixed above the centre of the tank, guides the cable so that it hangs clear and in a position to be easily manipulated by the gang of men, who gradually appear visible below as one becomes accustomed to the dim light shed by a few ship's lanterns hung around the sides of the tank. In the centre of the tank is a large iron truncated cone, which forms the eye of the coil of cable, and which, being hollow, also serves as a receptacle for perishable stores or fresh water for the consumption of the ship's company. The cable is arranged in flat coils occupying the whole space between the cone and the side of the tank; each coil is technically known as a "flake." In order to prevent one turn of the cable adhering to either of its neighbors, and thus producing a "foul," or a skein of several turns of cable coming up together when paying out, the cable is freely treated with whitewash to counteract the natural stickiness of the pitch-like exterior compound; as an additional precaution, boards are placed at intervals over each completed flake, thus obviating the risk of a "foul flake."

The whole scene, to an unaccustomed observer, possesses a weird, uncanny air; the gloomy cavernous tank, the lithe black cable, writhing

and swishing around with a ceaseless serpentine motion, the ghostly
figures of the men, who, viewed by the dim and fitful yellow light
below, seem like creatures of another world ; and to heighten the un-
earthly effect, a sort of gruff incantation, echoing and reverberating as
it ascends from the gigantic caldron, assails the ear and accentuates the
general resemblance to some séance of the black arts on a large scale ;
until, by listening intently, the mysterious notes are found to resolve
themselves into a chorus in vogue with sailors all the world over, but
peculiarly appropriate among such surroundings.

> " Heigho ! Roll the man down ! "
> " Heigho ! Roll the man down ! "
> " *Give* a man time to roll the man down ! "

The ships were loaded, the cable was all coiled snugly down in the
tanks, batteries, instruments, and stores were all stowed away, and on
the date appointed for sailing, which turned out to be a glorious
September day, we sped through the green fields of "the garden of
England," down to Greenhithe, where the two ships composing the
expedition were lying at anchor, only awaiting the final operation of
"swinging ship," and the arrival of the numerous staff of engineers and
electricians, who generally join the ships at the last moment. Our
train discharged quite a number of fellow-voyagers, some of them
accompanied by their friends. A turn of the road brought the river in
view, and right before us were the two good ships in which our principal interests were to centre for the next few weeks. They were looking
their very best; yards squared, rigging taut and trim, bunting flying
gayly in the autumn breeze; the blue peter at the fore, a few whiffs of
steam escaping from the waste-pipe, and a thin haze of smoke ascending
from the smoke-stacks, indicated that all was in readiness for departure.
At the landing-stage we found the ship's gig awaiting us, and in a few
moments we were standing on the deck of the *Dalmatia*, the flag-ship
of the expedition, as indicated by the swallow-tailed house-flag flying
at the main, which signified that we carried the commodore of the
squadron, in the person of the engineer-in-chief of the expedition.

The ship was in spick and span order, the deck clean and white, brass-work shining like gold, ropes coiled neatly away, wood and iron redolent of fresh paint and varnish; and were it not for the absence of guns and the very evident presence of the cable machinery which on all sides arrests the attention, we might have fancied ourselves on board some man-o'-war commanded by a strict martinet.

The operation of "swinging ship" was concluded, the boats were hoisted up to the davits, the accommodation-ladder hauled up and lashed securely to the rigging; the steam-winch was working heavily, and in a few minutes the anchor was weighed and we were steaming down the river. When we had the ship to ourselves, all the visitors having departed, the first thing to be done was to make a tour of inspection and gain some insight into the functions of the masses of heavy machinery which occupied the greater part of the deck from stem to stern. Starting from the bow we first observed the "bow sheave," a large iron pulley, deeply grooved, which projects out over the cutwater and serves to guide the cable in-board when the ship is engaged in "picking-up," a term which explains itself. The next prominent object was the dynamometer, a large iron sheave or pulley mounted on a frame, arranged so as to slide up and down, with a range of several feet, in a tall iron support; the wheel being balanced by weights, when the cable or grappling-rope is passed underneath, it indicates, by means of a pointer which passes in front of a graduated scale on the face of the iron support, the strain upon the rope or cable. Next we inspected the picking-up gear, consisting of a huge iron drum some six feet in diameter, worked by a powerful horizontal engine. Passing aft, we came to the paying-out gear, almost a replica of what we had already seen, except that the engines connected with the paying-out drum were of a lighter type than those forward, and that there were more appliances for holding the cable when it should be necessary, for any reason, to stop paying out.

The life on board a cable-ship is, as I have said, a thing of itself, differing widely from that of any other of the floating homes which at

all moments are ploughing the seas. This we soon found out as we
commenced to settle down and become familiar with our surroundings.

We were not on board a passenger steamer, because there were no
passengers of either sex : neither were we on a man-o'-war—we had no
big guns and no stern discipline. This latter element, however, was

Paying-out Gear, from Chart House.

not entirely absent on the *Dalmatia ;* every man on board had a certain
position and certain work to do, and all the members of the staff wore
uniforms similar to those of the ship's officers, the rank of each one
being denoted by the number of stripes on his sleeve. The engineer-in-
chief was the head of the whole expedition, and had entire charge of all
the operations, and the ships were navigated according to his instruc-

tions. Immediately after him ranked the captain of the ship, and the engineers and electricians of the cable staff, and the ship's officers and engineers followed in due order, according to their functions and standing in the company's service. Our party in the saloon also comprised two Spanish officials, who represented their government at all the operations of the expedition.

Cable engineers are naturally great travellers, and among our party of some twenty odd, a large proportion had visited almost every part of the world, and could relate many a good story of their varied experiences and give us much interesting information about foreign lands. Conversation in the saloon was carried on in at least three languages—English, French, and Spanish.

As our voyage was to be a very short one before we reached the port where we were to commence operations, little time was devoted to the amusements which while away the long hours on an extended trip. Everybody on board was busy preparing for the work in perspective. Here was a group of engineers conning over charts, studying the proposed track for the cable, and discussing the knotty point of selecting a suitable spot for landing the shore-end. A little further on, the paymaster, surrounded by papers, writing up his "log," and near by the hydrographer, preparing a large chart which takes in all the ground to be covered by the entire system of cables. In the testing-room, the electrician would explain the functions of the glittering instruments of ebonite and brass with which he was making a test on the cable in the tanks below. The only visible demonstration of what was being done was to be found in the movements of a little spot of light, which would be deflected from zero on a horizontal scale, and finally come to rest several hundred degrees to one side, as the assistant allowed the electric current to pass through the reflecting galvanometer. If the spot of light were to make sudden kicks or fly off the scale, the existence of something wrong would be revealed, perhaps a fault in the cable. But faults rarely develop on board ship, because the cable is perfect when it leaves the factory. In the ship's tanks it is kept cool by being always submerged in water, and as yet it has been subjected to no severe strain. When the time comes for paying-out, and the cable is straightened and has to bear a strain of several tons as it leaves the ship's stern, then any slight imperfection will be revealed; and although it may consist merely of a minute bubble of air which has burst and made a puncture in the gutta-percha into which you could not introduce a fine hair; although it may be only a crack so imperceptible that it would not admit of the insertion of the corner of a cigarette-paper, yet the current would escape, and, like the insignificant stream which trickles over a dam, would gradually widen the breach, until the cable was electrically " broken down," and entirely useless for communication.

Pondering over the watchful skill which manufactures hundreds, and even thousands, of miles of this slender cord with such widely different materials as iron, steel, hemp, gutta-percha, and copper, and triumphantly attains a degree of perfection which necessitates the exclusion of even such minute flaws and imperfections as would pass unnoticed in almost any other branch of industry, we dived down below to the main deck and spent an instructive half-hour inspecting the huge iron buoys, grappling-ropes and irons, mooring-chains and anchors, and other paraphernalia which the cable hands were busily painting, splicing, and overhauling generally in order to prepare them for use. On deck the same activity was to be seen; the heavy cable machinery was being examined and tried, to insure all being fit for action, and at the stern a small machine was being fitted up and got into place; this was the sounding machine, with which we shall shortly become more intimately acquainted.

The dreaded Bay of Biscay was crossed without undue pitching and tossing; for once its troublous waters were comparatively calm. In due course, one fine September morning, we steamed into Cadiz Bay. The scene is a beautiful one. On one side the bright, clean-looking little town almost entirely surrounded by the sea; on the other, some eight miles across the bay, the old town of Puerto Santa Maria. We were delayed a few days while the necessary formalities as to landing instru-

ments and stores, and other kindred questions, were gone through.
Some difficulty was also found in selecting a suitable landing-place for
the cable. Cadiz is surrounded by rocks, and also by currents. Rocks
are undesirable in the vicinity of a cable under any circumstances, but
rocks and currents combined arouse a feeling of unconquerable horror
and aversion in the mind of an experienced cable engineer. Finally, one
afternoon, when we had been at anchor in Cadiz Bay some three or four
days, orders were given for both ships to weigh anchor, and we found
that it had been decided to land the shore-end on a sandy beach at the
far side of the bay, near Puerto Santa Maria; the connection with
Cadiz town to be afterward made by means of a short cable skirting the
anchorage in the bay. Thus the main cable would be safe from damage
by rocks and currents, or by ships' anchors, and if the bay cable should
be broken at any time by either of these causes, communication could
always be maintained from the landing-place in the main line.

We steamed off and anchored as near in-shore as we could get,
opposite the spot intended for the landing-place. All was now activity
on board. No sooner were we at anchor than a couple of boats were
despatched for the beach, with a party of men and the necessary tools
and implements for use on shore. On board, both picking-up and paying-out gear were being made ready for action, as they both played
their part in landing the shore-end; huge coils of rope and a number
of collapsed air-balloons made their appearance from below. These balloons were inflated with air to their full diameter of some three or four
feet, and the quarter-deck of the *Dalmatia* began to assume the appearance of a giant's toy-shop. Meanwhile the shore party had firmly
anchored to the beach two large " spider-sheaves," or skeleton iron pulleys. These were placed some two or three hundred yards apart, forming two angles of a parallelogram, of which the bow and stern sheaves
of the ship made the other two. A rope was now carried from the stern
of the ship to the shore, and, passing round both spider-sheaves, brought
back to the ship and taken over the bow sheave to the picking-up gear.
The cable was made fast to the rope and paid out slowly over the

Landing the Shore-end.

stern, the picking-up gear meanwhile heaving-in on the other end of the rope, and so hauling the cable gradually ashore. The rope was wound four or five times round the big drum of the picking-up gear, steam was turned on, and the drum, rumbling and reverberating, hauled the rope in; aft, the cable was wound four or five times round the paying-out drum, also revolved by steam in order to ease the strain, which, with about a mile of rope out between the ship's stern and her bow, is something considerable. As the cable leaves the stern, the *raison d'être* of the air-balloons becomes apparent. At intervals of about fifteen or sixteen yards one is securely lashed to the cable, and in this way the cable is floated from the ship to the shore, and not dragged along the bottom to run the risk of being damaged by rocks. Another advantage is that, if the cable is sagged by a cross current or tide, it can readily be straightened by stopping the paying-out, and heaving-in at the bows.

So far all had gone swimmingly, and our first bit of cable was over the stern and fairly in the water, and we felt that the work of the expedition was begun in earnest.

However, interruption came from an unexpected quarter. The Spanish littoral is dotted around with coast-guard stations, the special mission of whose occupants (who are called *carabineros*) is the prevention of smuggling. We had no permission to land tools of any sort, much less a cable, and as we happened to pitch upon a spot close to a coast-guard station, the *carabineros*, alarmed at the sight of so many strange implements, came off in hot haste to order us to put a stop to our unlawful proceedings. It was explained to them that the cable was for the Spanish Government, and that everything had been arranged with the authorities in Cadiz; but they were obdurate, and, having received no instructions, were bent upon vindicating their authority. Your true Spanish official is nothing if he is not dictatorial, and the lower his rank the more authoritative he becomes. Diplomacy was then resorted to, and proved successful. The *carabineros* were assured that their demands should be complied with, and one of our best Spanish scholars was deputed to show them over the ship, *down below*. While they were being thus entertained (the contents of the chief-steward's bar formed no unattractive feature of the entertainment, and served to prolong it considerably), operations were continued, and by the time the *carabineros* came on deck again, a long line of balloons could be seen bobbing gayly on the water, all the way from the ship to the shore, and the end of the cable was safely on the beach. During the operation of landing the shore-end, communication was maintained between the party on shore and those on board by means of flag-signalling, a small hand-flag being employed to send messages in the Morse code. As soon as there was enough cable on the beach to reach to the site selected for the cable-hut, " Enough cable on shore," was signalled to the ship, and paying-out was at once stopped. The long rope was detached from the cable and rapidly hauled on board by the picking-up gear, boats were despatched to remove the balloon buoys from the cable and bring them back to the ship, while the shore party busied themselves in burying the cable on the beach and collecting the tools.

By this time it was nearly dark and flag-signalling had to be exchanged for flash-lamps, by which the *Dalmatia* signalled to the shore party to take all gear to the *Cosmopolitan*, as she was about to start paying-out seaward. All being made fast on shore and the last balloon buoy having been removed, we weighed anchor and set on slowly toward the open sea.

The cable now needed no steam power to help it out of the ship; on the contrary, it ran out freely of its own accord, and it was necessary to apply the brakes to the paying-out drum to prevent the cable running out too fast. It was astonishing to see the great, heavy, iron-bound cable, a single yard of which would weigh over ten pounds, come swishing round the tank, up on deck and over pulleys and guides, take four or five turns round a drum six feet in diameter, bob under the dynamometer, and up over the stern-sheave, and finally dive into the water with all the ease, grace, and pliability with which a silken cord might go through the same performance.

One striking thing in cable operations is the hearty will with which every one works, and the extreme anxiety evidenced on all sides for the welfare and safety of the cable. I have seen the engineer-in-chief, during the landing of a shore-end, up to his waist in the surf, cutting the lashings which secure the balloon buoys to the cable; and on another occasion, when, the ship being hove-to, the cable had got foul of the propeller, the chief of the expedition, after passing word to the ship's engineers not to move the engines, took a header into the water, and, holding on to a blade of the propeller, succeeded in freeing the cable. to the great relief of everybody on board, as all efforts from above had failed to dislodge it and a rupture seemed unavoidable.

During paying-out a test is always kept on the cable from the electricians' headquarters, the testing-room. Before the cable left the ship the end was carefully sealed by softening the gutta-percha and drawing it over the copper conductor; the cable was then charged with an electric current through the end on board, the current also passing through the galvanometer. We paid a visit to the testing-room and found by the steady deflection of the spot of light on the scale that the cable was sound and perfect.

The scene on deck is novel and interesting. The quarter-deck is brilliantly illuminated by electric light, which throws the mass of moving machinery and the figures of the men into bold relief; the big drum rumbles, and the pulleys and sheaves whir as the cable swishes over them, scattering whitewash in all directions. Every now and then a voice rings out announcing the number of revolutions of the drum, or word is passed up from the tank, couched in strange terms which we are only just beginning to understand. We have been paying-out for about two hours, when warning comes from the tank that only forty-five turns remain of the piece of cable which it was decided to pay out; the ship's engines are slowed down, and a few minutes later stopped altogether. A huge red iron buoy is in readiness, lashed to the mizzen rigging; paying-out is stopped and the cable made fast close to the stern sheave, the turns are taken off the drum, the cable is cut, and the extremity of the core sealed; the cable end is then secured to the moorings of the buoy, which consist of two heavy mushroom-anchors attached to the buoy by a length of stout iron chain. The lashings which hold the cable at the stern sheave are then removed, and the cable end is dropped overboard with the buoy-moorings; the chain rattles out with an appalling noise, above which a stentorian " Let go!" is heard, whereupon the buoy is released, and, dropping with a splash into the water, floats gayly off, dancing in the rays of the electric light. There the buoy will remain, securely anchored by its moorings, until the *Dalmatia* returns from the Canaries paying-out the main cable; the end of the piece we have just buoyed will then be brought on board and spliced on to the main cable, thus making it complete.

A Cable Buoy.

As we set on full speed for our anchorage every one on board felt that the work of the expedition had been successfully begun. An air of contentment prevailed on all sides; at dinner the health of the cable was drunk with due solemnity, and afterward an impromptu smoking-concert was held on deck.

On the following day, our business at Cadiz having been completed for the present, the expedition put to sea *en route* for the Canaries. The *Cosmopolitan* steamed out first, saluting the *Dalmatia* as she passed by dipping her ensign, to which we responded with three cheers, and a few hours later we followed suit.

The programme to be carried out by the two vessels was as follows: The *Cosmopolitan* was to make a zigzag course to the Canaries, taking short slants east and west of the proposed route of the cable, and sounding at intervals; the *Dalmatia* was to proceed in the same manner, except that her zigzags were to be longer and at a different angle to those of the *Cosmopolitan.* In this way it was hoped that a thorough survey would be made of the ocean depths between Cadiz and the Canaries, and a safe route selected for the cable. At Cadiz, our scientific staff had been augmented by the arrival on board of a distinguished chemist and naturalist, who accompanied the famous *Challenger* expedition, and who, therefore, was an authority on the subject of ocean surveys, and took a vast interest in all such matters. This gentleman was prepared to analyze and tell us all about the constitution and properties of as many samples of "bottom" as we could obtain for him, and he has since produced some remarkably interesting papers of high scientific value, embodying the results of the immense amount of work performed by the expedition.

By the time we got clear of Cadiz harbor the *Cosmopolitan* was "hull down," and we saw no more of her till we met in Grand Canary. The course of the *Dalmatia* was shaped for the Straits of Gibraltar, and soon after leaving Cadiz we took our first sounding. The little machine which then came into action, and played a prominent part in the work of the next few weeks, is worthy of a little attention, both on account of its simplicity and because of the amount of good work that it performs in a rapid and trustworthy manner. The sounding machine consists mainly of a light iron drum or spool, upon which are wound several thousand fathoms of steel pianoforte wire; to the wire is attached a sinker which is provided with a receptacle at the lower extremity for securing a specimen of the bottom. When the wire is being paid out the drum projects over the ship's stern, and for hauling in it is run in-board a few feet and connected to a small steam engine, which makes short work of winding up the wire and bringing the sinker to the surface. Besides the ordinary sinker there is a whole battery of other apparatus, such as sinkers with weights which are detached automatically on reaching the bottom, leaving only the tube to be brought up; thermometers which register the temperature of the water at different depths; tubes constructed to obtain samples of water from the bottom, and so on *ad infinitum.*

Our first piece of scientific work was a survey of the "Gut," as the entrance to the Straits of Gibraltar is commonly called by mariners. This was slightly out of our strict programme, but served to get our hands in for more important operations to follow.

Having spent nearly three days in this interesting work, during which time we obtained a quantity of new and valuable information as

Sounding Machine.

to the formation of the bank at the entrance to the Mediterranean, we started out seaward, and rapidly got into deep water. Here the sounding machine showed to great advantage. In olden times, when hemp lines were used for sounding, it was necessary to employ a weight of about four hundred and fifty pounds to keep the line vertical, and about three hours were occupied in taking a sounding in a depth of two thousand fathoms. With steel wire we used a sinker of only fifty pounds, which in twenty-two minutes reached bottom at a depth of a little over two thousand fathoms; there was a delay of a few minutes in detaching the weight and in connecting the drum to the engine to wind in. The weight was detached automatically, the wire by which it was suspended to the tube being cut through by a hinged knife on the head of the tube at the moment when strain was applied to wind in; the weight was thus left on the bottom and the tube alone brought to the surface. In this way there is very little strain on the wire, and consequently but slight risk of breakage. The little engine commenced to buzz away, and in forty-eight minutes from the time of letting go the tube was on board again, and the ship proceeded on her course. We all crowded round to examine the little instrument which had made its venturesome descent through some two and a half miles of blue water. General satisfaction was caused by the fact that the specimen obtained was one of *globigerina* ooze, which consists of myriads of tiny shells of carbonate of lime. The existence of this ooze denotes the entire absence of currents, and the ooze itself forms a soft, yielding bed into which the cable would sink luxuriously, and might rest undisturbed to the end of time.

About every four hours we stopped to take a sounding, and the results were almost invariably satisfactory. Occasionally a sounding was spoiled by the wire kinking and breaking, the consequence being the loss of the tube and a certain amount of wire; but so carefully were the operations conducted that this was a very rare occurrence. Deep-sea sounding is very interesting work, but it is a trifle annoying sometimes to hear the engine-room gong sound, and have to leave a good hand at cards and rush up on deck, especially if the weather is rough, when the whole sounding party stands a chance of getting a good drenching from a "poop sea."

One night we were astonished by the sinker stopping at about one thousand two hundred fathoms, when it ought to have gone nearly twice as deep. It was at once suspected that we were in the neighborhood of a bank. A sounding was taken three miles further on and showed deeper water, so we retraced our course eight miles; here we got only eight

13

hundred fathoms.   Expectancy then ran high, and it was fully justified
when, two miles further back, the sinker stopped at four hundred and
fourteen fathoms; but the crowning event occurred at the next dip,
after another run of two miles.   Here, to our surprise and delight, the
sinker brought up at sixty-six fathoms !   There was immense excite-

Paying-out Gear. From Stern Baulks.

ment on board, as it was obvious that we had pitched upon a bank, or
rather a mountain, of startling proportions, perhaps the lost island of
Atlantis itself.   As this submarine mountain lay close to the proposed
line of the cable, it was necessary to make a thorough survey, and two
days were spent in doing this.   A mark-buoy was put down to work

by, and numerous soundings were taken in all directions so as to clearly define the limits of the bank. The shoalest water found was forty-nine fathoms, and half a mile distant two hundred and thirty fathoms were obtained, showing a steep slope. When the buoy, which was moored in one hundred and seventy-five fathoms, was taken up, the mooring rope was found to be nearly chafed through seventy-five fathoms from the bottom. This showed that the bank must rise almost precipitously, and that there exists a wall of about four hundred and fifty feet in height. A very curious effect observed was a long ripple on the calm sea, apparently caused by the ground-swell breaking on the edge of the bank.

Nothing further of an exciting nature happened during the soundings, and after one more zigzag our course was shaped for Grand Canary, our rendezvous with the *Cosmopolitan*. The *Cosmopolitan* had made no such interesting discoveries as had fallen to our lot, and having been awaiting our arrival several days, those on board finally became alarmed at our delay and started out to look for the *Dalmatia*. We met the night before our ship was due to arrive at Canary, and rockets being fired, the two steamers recognized each other, and a conversation was kept up by means of the steam-whistles, the Morse code adapting itself as well to this method of signalling as to any of the many others in daily use.

The following morning both ships were at anchor in the harbor of Las Palmas, the capital of Grand Canary. During the next week or two we visited the different islands, taking soundings between them and spending a few days at each port. Receptions were given on board to which the authorities and principal inhabitants were invited, and all the wonders of the ships were explained to them. Everywhere the greatest enthusiasm was displayed, as the natives looked upon the establishment of telegraphic communication as a great step in putting them in touch with the civilized world. Public rejoicings and *fêtes* were the order of the day. At Las Palmas a ball was given to the officers and staff of the expedition, and (considering that we were in such an out-of-the-way place) we were fairly astonished at the scale of magnificence on which the entertainment was carried out, and at the dresses and jewels of the ladies, while not a few members of the staff were considerably smitten with the personal charms of their partners; but, unfortunately, with but few exceptions, they could not exchange five words with them. At Teneriffe the chiefs of the expedition were escorted through the streets by a band of music and an immense crowd, and at La Palma, the western island of the group, the ships were serenaded, the town was *en fête* and decorated with triumphal arches, and another ball was given. Altogether, we were the heroes of the day throughout the Canaries.

It was decided to lay the cable between Teneriffe and La Palma first, and the necessary soundings having been taken, both ships steamed round Teneriffe one fine November evening, and came to anchor off Garachico, a little village on the southwest coast of Teneriffe. Here it was proposed to land the cable, the connection between Garachico and Santa Cruz, the capital of Teneriffe, to be afterward made by a landline across the island.

At Garachico we spent several days. The coast being barren and rocky, considerable difficulty was experienced in finding a suitable landing-place for the shore-end. Finally a spot was selected, and the shore party signalled that they had engaged a team of oxen to haul the end on shore, as the bad ground rendered it unadvisable to employ the usual method of working the whole operation from the ship. Everything went well and the end was soon successfully landed, and all being made fast on shore, the *Dalmatia* paid out about a mile of cable seaward; then cut and buoyed the end in the same manner as at Cadiz.

The next few days were occupied in erecting the cable-hut (a small structure of galvanized iron about twelve feet square), in fitting up the testing instruments in the hut, and in transferring a few miles of heavy cable from the *Cosmopolitan* to the *Dalmatia*. Finally all operations at Garachico were completed, and early one morning we started for the buoy and picked it up, and with it the end of the cable secured to the buoy moorings. The cable end was brought on board and spliced to the cable in the tank from which it was intended to pay out. The splice is always an interesting operation to watch. First the jointer and his assistant go to work and nimbly and rapidly join and solder the ends of the copper conductor, and then cover it over with sticky black compound and gutta-percha sheet, producing a homogeneous joint but little larger than the machine-made core, and every bit as impervious to

Cable-hut at Shore-end.

the action of the water. The joint is tested by the electricians to make sure that it is sound and perfect, and this being ascertained, the cable hands at once go to work on the splice ; and it is surprising to observe how skilfully they manipulate the stiff iron wires, first carefully wrapping the core with its protective hemp covering, then laying on the armor wires and butting them together, and finally winding over the whole length of the splice a stout cord of spun yarn.

The splice was finished and we started paying-out, slowly at first, but with gradually increasing speed, until deep water was reached and the light deep-sea cable went whizzing through the machinery at the rate of seven or eight knots an hour. Now we were at work in earnest. One of the engineering staff was in charge of the quarter-deck, keeping a watchful eye on the dynamometer and the indicator on the paying-out drum ; by the former he knew the strain on the cable, and by the latter the amount of cable paid out ; of these data an assistant was continually taking notes. In the testing-room we found that a careful watch was being kept on the electrical conditions of the cable. The sensitive spot of light was doing its duty both here and in the cable-hut, and the electricians on shore exchanged signals every few minutes with those on the ship. Thus both the mechanical and electrical behavior of the cable were continually under such scrupulous and accurate observation that it was impossible for anything to go wrong without those in charge being at once aware of it. The ship steamed steadily ahead and everything worked as smoothly as clock-work ; coil after coil of the cable unwound from the tank, glided over pulleys and through troughs, wound around the swiftly revolving paying-out drum, dived under the wheel of the dynamometer and over the stern sheave, and trailed away after the ship until, a good many yards astern, it silently dipped into the water to seek its final resting-place in the motionless depths.

As darkness came on the arc lamp was lighted, and with the aid of its brilliant rays work was done as easily as during the day-time. Toward midnight we approached La Palma, and the *Cosmopolitan* steamed ahead to show us a good position for buoying the end, which operation was necessary, as the La Palma shore-end had yet to be laid. Gradually our speed was slowed down ; the electrician on duty in the testing-room informed those in the hut at Garachico that we were about to cut the cable and buoy the end, and immediately afterward, as the ship had come to a standstill, the cable was made fast, the turns were taken of the paying-out drum, the executioner advanced with his axe and severed the cable, the wounds to its centre nerve were healed up by means of a spirit-lamp, it was fastened securely to the moorings of the buoy, and in a few minutes cable, moorings, and buoy were all overboard and we steamed off for port.

The next day the *Cosmopolitan* took up the work and met with ill-luck, which proved to be only the commencement of a series of disasters. To begin with, while the cable-hut and tools were being landed, one of the boats was capsized by the surf, the contents scattered broadcast, and a man imprisoned under the overturned boat. This unfortunate was, however, quickly rescued by his companions and equally quickly resuscitated, being more frightened than hurt. The shore-end was successfully landed, and, as night was coming on, the *Cosmopolitan* started to pay out toward the buoy put down the previous night; the buoy was picked up and the mooring-rope taken to the picking-up drum, which at once commenced to heave in; but after a few turns a sudden diminution of the strain on the rope showed that it had parted, and the end of the cable was lost! There was nothing to be done but buoy the end of the short length just paid out and return to port, as it was too late to attempt to grapple for the lost cable.

For the next two or three days the weather was so bad that nothing could be done; but finally, when everybody's patience was thoroughly exhausted, wind and sea moderated sufficiently for us to set to work. A grapnel was lowered over the bows by means of a long rope, the end of which was taken under the dynamometer to the picking-up drum. The dynamometer serves in this case to show when the grappling-iron hooks the cable, as it at once indicates the increased strain on the rope. We steamed slowly back and forth across the course of the cable, and made four or five unsuccessful drags. Once we hooked the cable, but only succeeded in bringing up a loose piece, as it parted further seaward. The scene on board now is very different to a few days back, when paying-out was going on so smoothly. All the machinery on the quarter-deck is motionless and deserted; in the testing-room the active little spot of light is extinguished and the place wears an untenanted air: interest is concentrated forward, where the engineers watch every

rise and fall of the pointer on the dynamometer with acute anxiety.
Electricians and others on board who find their occupation gone, hang
about, listless and dejected, and a general air of discontent reigns. We
are grappling in deep water, and, as is evident by the jerky action of
the dynamometer, on rocky ground; but finally, after a long and weary
day, a steady strain is observed, the picking-up drum is set to work, and
after a vast amount of laborious puffing and rumbling, shortly before
midnight the grapnel arrives at the bows with the cable securely suspended across two of its prongs! At once all is activity on board.
The testing-room brightens up and the spot of light shines cheerfully
once more. The cable is cut and handed over to the electricians to be
tested. Very shortly the verdict is delivered to the effect that it is in
perfect condition, and at once the operation of splicing it to a new length
of cable in one of the tanks is commenced; this concluded, we start
paying-out, and all goes well until we reach the buoy on the shore-end.

Here a double disaster occurred; the experience of the *Cosmopolitan*
was repeated, as the moorings broke shortly after we commenced heaving-in. It was then necessary to pick up a short length of the cable we
had just laid, so as to cut and buoy further out.

While this was going on we dropped into the testing-room to see
that matters were all right there, and scarcely had we commenced to
watch the spot of light, when it quivered, oscillated, and finally darted
off the scale. Something was wrong, and we made for the deck, where
our suspicions were confirmed ; the cable had broken, and a few minutes
later we were all gazing mournfully at the jagged end—a mere bunch
of tangled wires and hemp! Both ends were now lost, and there was
nothing for it but to start grappling again. Drag after drag did we
make with the same lack of success; occasionally the strain went up
with a rush as the grapnel clutched a rock, only to decrease with equal
suddenness as the rock gave way and the grapnel flew off. Our spirits
rose and fell with the pointer of the dynamómeter, and when it only
indicated the normal strain of the rope and grappling-iron, we all sank,
mentally speaking, far below zero.

This sort of thing went on all day. At 12 P.M. the grapnel was at the bows, but no cable, so work was suspended for the night and every one turned in for a well-earned rest. The following day our luck changed. The cable was hooked at the first drag and brought safely on board; the tests showed that it was still perfect, and the splicing and paying-out were proceeded with in due course. Meanwhile the *Cosmopolitan* had grappled and rebuoyed the other lost end, so we had no more difficulties to encounter. While paying-out, the submarine crater over which we evidently had been working, and which had given us so much trouble, was carefully avoided by taking a circuitous route. The buoy was soon reached and the other end hauled on board. Both cables were carefully tested and pronounced to be perfect, the final splice was made, and with three hearty cheers the completed cable was lowered overboard.

Finis coronat opus. Our first complete section was finished, and Teneriffe and La Palma were in telegraphic communication with each other.

The rest of the work among the islands was carried out without a hitch of any sort, the long cable from Teneriffe to Cadiz being left to the last. This was, of course, a matter of several days, and may be taken as a good example of the routine on board when laying a long cable. Mile after mile of cable goes steadily out; the machinery whirs and revolves as if it never would stop, the spot of light in the testing-room behaves with perfect propriety, and only oscillates once every five minutes, when those on board exchange a signal with the man on watch in the cable-hut at Teneriffe. Every four hours tired engineers and electricians go below and take their share of refreshment and rest, as sleepy substitutes come on deck to take their places. One startling incident relieves the monotony of this prosperous state of affairs. On the third night out, the eccentric behavior of the dynamometer, indicating a varying strain, shows signs of an irregular bottom. At the same moment the *Cosmopolitan*, engaged in taking soundings a few miles ahead, is seen to fire a rocket. Shoal water is immediately suspected, and the

Dalmatia is put full speed astern and cable paid out freely. It was found that the *Dalmatia's* course lay directly across a bank with only eighty-four fathoms of water on top, and nothing but the prompt way in which the situation was grasped by the engineer on watch averted an accident; for if paying-out had been continued at full speed, the cable would have festooned from the edge of the bank and most infallibly been broken.

The foregoing narrative of a cable-laying expedition is a typical description of the manner in which the great work of lessening the separation set up between continent and continent by the trackless ocean is carried out. Nowadays it is not the good fortune of all cable expeditions to open up new ground and be welcomed and feasted by the natives, as much of the cable work which is being constantly carried on in all parts of the world consists of the renewing, duplication, or triplication of existing lines; and the laying of a new cable has come to be so much a matter of course that such an event arouses the merest spark of passing interest, although books which have become classical were published chronicling the progress of the early Atlantic cable expeditions.

The reader has taken a glance at the manufacture of the submarine cable of to-day; he has seen how the ocean depths are surveyed almost with as much care as the land for a new railroad; he has watched the landing of a shore-end, and has seen the deep-sea cable trailing steadily out into blue water; he has participated in the joy and enthusiasm of dropping overboard a final splice, and in the disappointments and anxiety attendant on grappling for a broken cable on rocky bottom. Altogether he has made a fair acquaintance with life on board a cable-ship; and if he can point out any other branch of electrical work equally interesting and fascinating, I should much like to know which he would select.

ELECTRICITY IN NAVAL WARFARE.*

By LIEUTENANT W. S. HUGHES, U.S.N.

THE FIRST NAVAL VESSEL LIGHTED BY ELECTRICITY—SATISFACTORY TEST ON A THREE-YEARS' CRUISE—THE SEARCH-LIGHT—HOW CONSTRUCTED, MOUNTED, AND OPERATED—STATIONARY SEARCH-LIGHTS—SIGNALLING OVER GREAT DISTANCES—THE DEVICE OF THE UNITED STATES NAVY FOR MORSE SIGNALS—SAFE RUNNING LIGHTS—SUBMARINE ILLUMINATION—THE ELECTRIC TORPEDO DETECTOR—GUN SIGHTS ILLUMINATED AT NIGHT—FIRING BIG CANNON BY ELECTRICITY—ELECTRIC LOGS—SMALL MOTORS ON MEN-OF-WAR—IMPORTANT SERVICE IN PROPELLING "FISH" TORPEDOES—APPLICATION OF ELECTRICITY TO DYNAMITE CRUISERS—AN ELECTRICAL RANGE FINDER—TELEGRAPHING AT SEA—ELECTRICAL STEERING GEAR.

THE extension of the applications of electricity has been nowhere more rapid or remarkable than on board ships-of-war. Only a little more than seven years have elapsed since the Navy Department fitted out the frigate *Trenton* with an electric incandescent plant, and, so far as is known, she was the first naval vessel to be lighted by electricity.

The experiment with the *Trenton* was watched with interest both at home and abroad. Serious apprehensions had been entertained that firing the ship's guns would break the glass bulbs, or destroy the carbon filaments of the lamps, and that other causes would operate against the successful employment of electricity on board a man-of-

---

* In the preparation of this chapter the writer has been greatly indebted to valuable reports made to the U. S. Office of Naval Intelligence by Lieutenant J. B. Murdock, U.S.N., and by Mr. S. Dana Greene, late Ensign, U.S.N.

war ; but the thorough test on the *Trenton*, lasting through a three-years' cruise, proved beyond question the fallacy of such fears.

Foreign countries were not slow to follow the footsteps of the United States, and the latter government gradually extended the system of electric lighting to other vessels of the navy.

It had been early discovered that the installation of an electric plant on board naval ships required a decided modification of the electrical appliances employed for similar purposes on shore. The limited space that could be spared on the crowded decks of a man-of-war demanded that engines and dynamos of a special and compact type should be used ; and these requirements were further complicated by the necessity, which soon developed, of providing naval vessels with two systems of illumination. Gradually, however, man's ingenuity overcame every obstacle—and how completely, the reader will not fail to appreciate by a glance at the wonderfully compact electrical machinery included among the illustrations of this chapter—until, at the present day, no modern-built vessel-of-war goes to sea without complete electrical plants, both "arc" and "incandescent," the latter principally for the purpose of. interior lighting, and the former for operating the so-called "search-light."

The name of the "search-light" suggests, to some extent, the chief purpose for which it is used. It is mainly employed in searching for an enemy. It consists of a powerful "arc" light, usually of about twenty-five thousand candle-power, contained in a metal cylinder about thirty inches long by twenty-four to thirty inches diameter. A good idea of its construction and general appearance may be obtained from the illustrations.

One end of the cylinder is closed by a silvered. concave reflecting lens ; and the carbon points of the lamp, as represented in the sectional view, are placed in such a position within the cylinder as to bring them in the focus of the lens. The opposite, or front, end of the cylinder is fitted with a glass door through which the beam of light passes. Other mechanical features are shown in the illustrations.

The whole apparatus is mounted on a pivot, so that it may be revolved around its centre—and the beam of light be thus thrown in any

Ship-of-war's Search-light, 25,000 Candle-power.

(From a photograph furnished by Messrs. Sautter, Lemonnier & Company, Paris.)

desired direction—while it admits, also, of elevation or depression from a horizontal position.

As ordinarily used on ships-of-war, the beam of light emerging from the cylinder is so concentrated that at a distance of a thousand yards from the ship it illuminates a path only about fifteen yards in width. Every search-light, however, is provided with arrangements for increasing the divergence of the rays of light, in order, when necessary, to illuminate a broader arc. Only one man is required to work the apparatus, and to facilitate its use it is always, on board ships, mounted in some position considerably elevated above the deck.

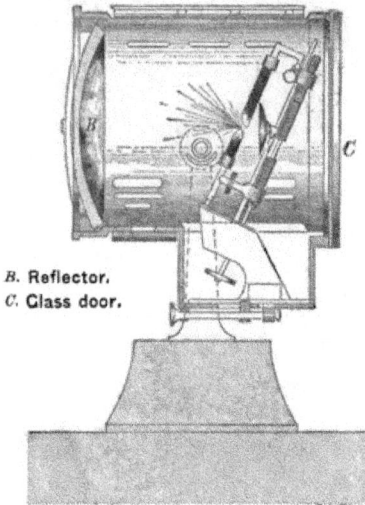

B. Reflector.
C. Glass door.

. Sectional View of Ship's Search-light.
(From a sketch by Ensign H. H. Eames, U.S.N.)

In time of war, ships at night would constantly sweep the surrounding waters with their search-lights, illuminating, in succession, every part of the circle around them, in order to detect the presence of an enemy; and the practical value of the light, in such cases, will be readily comprehended when it is understood that, with one of only twenty thousand candle-power, no difficulty is experienced in lighting up any object, such as an enemy's ship, at a distance of two and a half miles, thus rendering firing the guns at night as easy and accurate as by day.

The view of the *Empong*, one of those phenomenally swift little vessels called "torpedo-boats," shows its search-light mounted on top of the pilot-tower, just in front of the two smoke-pipes.

In the practical use of the search-light, by the method just explained, it has been found that, in order to afford sufficient time for a careful examination of the water's surface at points far removed from the ship, the beam of light must be revolved very slowly; and, in consequence, during a great proportion of the time any particular section of the water is in darkness. When it is remembered that it requires less than five minutes for modern torpedo-boats to pass over a distance of two miles, it may be easily conceived how, in the interval between two successive illuminations of the same spot, a little craft like the *Empong* might dart in and discharge its torpedoes.

With the object of averting such a danger, another method of illumi-

Yarrow Torpedo-boat "Empong," showing Search-light.

nating the surrounding space has been adopted on one or two ships of the French Navy, and on the Danish cruiser *Ivert*, where, instead of a single revolving light, a number of stationary search-lights are grouped together, each illuminating its proper section, thus keeping the ship continuously surrounded by an unbroken circle of light.

But the search-light is not only used to discover an enemy, and to keep the latter visible in firing at night; it has been, also, successfully employed in signalling messages where the distances over which they are to be sent are very great. One of the methods of using it for this

purpose is somewhat novel : the beam of light is simply flashed repeatedly against the clouds, lighting them up in each instance for a certain
number of seconds, according to a pre-arranged code of signals, the
letters of which are indicated by combinations of flashes of different
durations of time.   Messages are said to have been sent in this manner
between ships separated by a distance of sixty miles.   For shorter distances the ordinary incandescent lamps
have been frequently used.   Probably
one of the most successful systems with
these lamps is that now in use in the
navies of Italy and Germany.   Its essential feature consists in the successive
display of different combinations of red
and of white incandescent lamps of about
fifty candle-power each.   It is, of course,
necessary that the lamps be
exhibited in some elevated
position, such, for example,
as suspended from one of
the ship's yards, where they
will be plainly visible from
a considerable distance, and
that the electric current con-
nected with them be under the control of an operator at a key-board, in
order that any lamp may be lighted or extinguished at pleasure.   With
such an apparatus it will be readily understood how the letters of a
signal-code could be indicated, and a message rapidly transmitted.

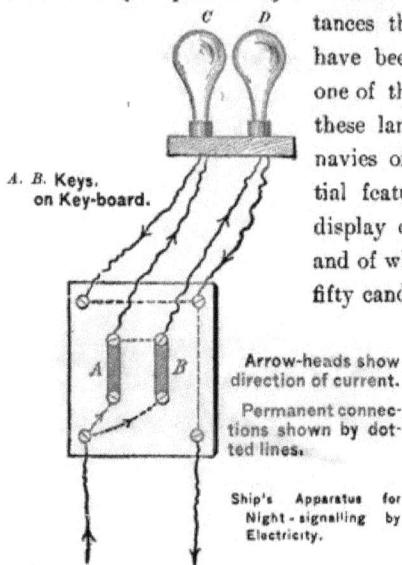

In the United States Navy a more simple plan, requiring only two
incandescent lamps, has been very successfully used in connection
with Morse's telegraphic code, in which, as is well known, the letters of
the alphabet are represented by combinations of "dots " and " dashes."
In signalling by this code the two lamps were controlled by keys, as in
the figure shown above.

Figure labels:
A. B. Keys. on Key-board.
Arrow-heads show direction of current.
Permanent connections shown by dotted lines.
Ship's Apparatus for Night-signalling by Electricity.

It will be observed, from an inspection of the electrical connections in the figure, that when the key A is pressed, the lamp C is alone ignited, or "flashed;" and that when B is pressed, both lamps are flashed simultaneously. To indicate a "dot" of the Morse code, only a single lamp is used; to denote a "dash," both are ignited together.

Every steamer displays, when under way at night, as doubtless most of our readers have observed, three lights of different colors *— green, red, and white. These "running lights," as they are technically called, are exhibited on the outside of the ship, in such positions as to indicate to other vessels the approximate direction in which she is moving. To avoid collisions, and loss of human life, it is of vital importance that these lights be never for a moment extinguished. With the aid of electricity the possibility of such an accident as the unexpected extinguishing of a "running light" at a critical moment will be reduced to a minimum, if not altogether eliminated, by an ingenious appliance in which the breaking or burning out of one incandescent lamp *automatically* completes a circuit and ignites another. Only a casual inspection of the figure will show how this is effected : A small electro-magnet, E, is placed in the circuit of the lamp 1; and, while the latter is burning, the attraction of the magnet keeps the armature D in the position shown by the dotted lines, with the circuit of lamp 2 broken. But, should the circuit of 1 be broken by the burning out or breaking of the lamp, or from any cause, the magnet will, of course, cease to attract the armature, and the latter is at once drawn back by a spring, S, to the position shown in the figure, thus

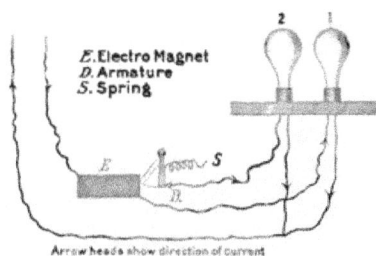

E. Electro Magnet
D. Armature
S. Spring

Arrow heads show direction of current

Arrangement of Electrical "Running Lights" for Ships.

* A green light on the starboard (or right) side, a red light on the port side, and a white light at the masthead.

14

completing the circuit and igniting lamp 2. The figure represents this as having already occurred.

Another application of the incandescent lamp has developed from a series of experiments made at the Naval Torpedo Station, Newport, R. I., with the object of ascertaining to what extent electricity could be utilized at night for lighting up water below the surface. In these experiments incandescent lamps of about one hundred and fifty candle-power were immersed to a depth of fifteen to twenty feet, with the result that sunken objects could be readily seen within the distance of a hundred feet from the light. The discovery thus made is of considerable importance from a naval point of view. Modern naval fleets must not only be prepared to rapidly remove the stationary torpedoes, or "fixed mines," with which every important port in time of war will be protected, but the ships themselves must be provided with a complete system of such weapons, ready to block up harbors in which an enemy's vessels may be discovered, or in which they themselves may have taken refuge. Much of this service, in the event of war, would be necessarily performed at night, and a part of it under water; hence, it may be certainly expected that modern navies will avail themselves of the valuable agent that the Newport experiments have placed in their hands. A portion of the force employed for such duty would be a body of trained divers; and not only will these men carry with them under the water an electric light, but the telephone has been recently added to their outfit, so that they may communicate constantly and easily with assistants above the surface.

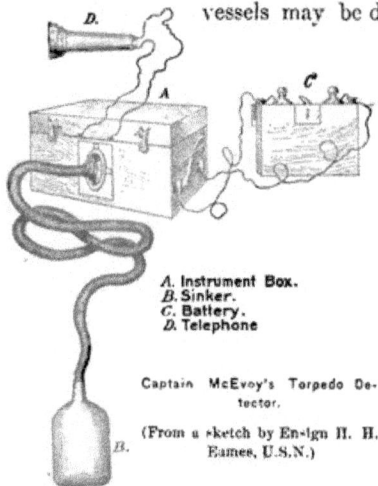

A. Instrument Box.
B. Sinker.
C. Battery.
D. Telephone

Captain McEvoy's Torpedo Detector.

(From a sketch by Ensign H. H. Eames, U.S.N.)

In this connection may be mentioned a remarkable electrical apparatus known as a "Torpedo Detector." As it name implies, it is designed for use in searching for hidden torpedoes, or mines, in harbors and channels. This appliance depends for its action upon the principle of the induction-balance. A representation of it is shown opposite. The box *A*, in the figure, contains a set of induction-coils, and a vibratory magneto-electric machine, with which a telephone is connected. The metal sinker *B* contains, also, a set of induction-coils, and is connected with *A* by a short electric cable. The small battery *C* has connections as shown in the figure, the circuit of which can be opened or closed by a key.

In practical use, searching for torpedoes, the box and battery are carried in a boat, and the metal sinker, to which the small cable is attached, is dragged overboard along the bottom of a channel or harbor where the presence of torpedoes is suspected. When the telephone is placed at the ear, no sound is audible until the sinker reaches the vicinity of some metal body, such as a torpedo; then a buzzing noise is heard, which becomes gradually louder as the torpedo is approached, and loudest when it is touched.

Electricity is employed on some of the vessels of the British Navy, and will probably be made use of on the new ships of the United States service, to render visible the sights of the guns when firing at night. As applied on the English iron-clad *Colossus*, one of the wires from a small Leclanché battery leads to the rear, or "breech" sight of the gun, and is there joined to a fine platinum wire running across the bottom of the "sight-notch." The platinum wire interposes just sufficient resistance to cause it to glow with heat while the current from the battery is passing through it. From the rear-sight the battery wire continues on to the front-sight of the gun, where it meets the second wire from the battery. Here the ends of the two wires are brought very near each other, at the apex of the sight, so that the electric sparks passing between them serve to mark its position. The practical utility of such an arrangement may be shown by a familiar illustration: Every one has observed, when looking from the window of a lighted room on a dark night, how difficult it is to distinguish any outside object beyond the distance of a few feet. The same effect is naturally experienced on board ship, and for this reason, among others, men-of-war, on going into battle at night, have every unnecessary light extinguished—thus rendering *some* plan of indicating the position of the guns' sights almost absolutely indispensable. That above described, the writer was in-

Ship-of-War using her Search-lights.

formed by an officer of the *Colossus*, has proved perfectly successful in actual practice.

On some of the larger European ships-of-war, and on a few United States vessels, arrangements have been made for firing the guns by electricity. This is usually accomplished by causing the current from an ordinary zinc-carbon battery to pass through what is known as an electric primer, inserted in the vent of the gun. The primer, as used in the United States service, is simply a quill tube nearly filled with small-grained powder, and containing a fine platinum wire wrapped with a wisp of gun-cotton. The platinum wire is connected with the wires of the electric battery. When the battery circuit is closed, the platinum becomes instantly white-hot, thus igniting the primer and firing the gun. The firing-key, with which the circuit may be closed at pleasure, is placed in the pilot-tower, or in some other position where it is directly under the control of the commanding officer, and the circuit is, of course, kept open until the instant of firing. The advantages claimed for such a system are many. Obviously, under some circumstances, as, for example, where it is desirable to concentrate the whole broadside upon a certain point, and to fire all the guns together, such an arrangement would have a great advantage over the ordinary method of depending upon the simultaneous action of the gunners. Again, it is no uncommon occurrence in battle for a ship's gun-decks to become so enveloped in smoke that the enemy cannot be seen by the men at the guns, and in such cases, with an electric system, the firing could be done by an officer clear of such an obstruction. Usually the electrical appliances admit, also, of the guns being fired singly and in succession; and for cases where the ship is rolling heavily from side to side, an "automatic circuit-closer" is sometimes employed, which, *after the regular firing-key has been pressed,* closes the circuit and discharges the guns the instant the vessel reaches an upright position.

Electric "logs," for measuring the speed of a ship, usually depend for their action upon the revolutions of a small screw-propeller, or "fan," attached to the end of a short electric cable, and trailed through the water astern of the vessel. The velocity with which the fan revolves at any given speed through the water has been previously determined by an actual trial. The wires of its cable are connected with a battery on board the ship, and its revolutions periodically open and close the circuit of the battery, which action is communicated by suitable mechanism to a register, or dial, so graduated as to show the corresponding speed in miles and tenths. While such an apparatus is valuable for the purposes of navigation—since it affords an accurate measure of the distance passed over by the ship in the *interval* between any two readings of the dial—it fails to show the speed of the vessel at the *moment of observation.*

This latter knowledge will be an important factor in future naval fights. The commander of a modern ship-of-war engaged in battle must be able to ascertain the rate at which his vessel is moving at any desired moment, in order to make allowance for the deflection of shot, and still more for that of torpedoes, due to the speed of the ship at the instant of their discharge. Hitherto, so far as the writer knows, only a single "log" has been designed which accurately furnishes this essential information. This instrument is an ingenious "speed indicator," the invention of Lieutenant Hogg, of the United States Navy, but of

Torpedo-boat Electrical Plant, now in use in the Navies of Italy, Spain, Austria, France, Russia, and Denmark.

which, since it depends upon no principle of electricity (but upon that of the vacuum-gauge), no more than the mere mention of its existence can be here made.

The question of introducing electric motors in ships-of-war, to take the place of the numerous small steam-engines now required on board for various purposes, is one that is occupying the serious attention of naval authorities. That such a substitution will be made in the near future seems now assured. An electric current is the most convenient method of transmitting power from one point to another, at a considerable distance, that has been yet discovered. It is also the most economical where the conditions, as on board a modern man-of-war, require the power to be conveyed from a single source to a number of different

points of application.   Yet, at the present writing, no use of an electric
motor is made in any of the great navies of the world.   It has been left
to the United States to take the initial steps toward its introduction, by
employing electricity to work one of the eight-inch guns of the steel
cruiser *Chicago*. This motor is supplied with the requisite electric current by connecting its wires with those from the dynamo used for lighting the ship.

Electricity has been the chief element in the development of those
remarkable inventions known as auto-mobile, or " fish," torpedoes.

Revolving Deck Tubes for Yarrow Torpedo-boats, showing electrical discharging apparatus.

They are, with few exceptions, steered, propelled, or exploded by an
electric current ; and in some all three of these applications are made.
No ship-of war is now regarded as efficient whose armament does not
include a number of torpedoes.   The " Whitehead " is the torpedo in
general use in the navies of Europe.   Briefly, it is cigar-shaped, from 9
to 19 feet long by 11 to 16 inches diameter, made of thin steel, propelled
by compressed air, and carries an explosive charge of gun-cotton ranging from 40 to 105 pounds.   In practical use the torpedo is discharged
at the enemy's vessel through a tube, which may be mounted on board
ship either above or below the surface, and upon reaching the water it
is propelled by its own engines.   The initial motion, or " discharge," is

effected by an electric fuse which ignites a very small charge of gun-powder placed in the tube behind the torpedo. The wires from the fuse are connected in the circuit of a battery, and lead to the pilot-tower, where the firing is accomplished by the usual means of a key-board under the control of an officer. The illustrations show the tubes in place on the deck of a torpedo-boat, and also represent them separately on a larger scale. It will be observed that they are mounted on a turn-table, so that they may be pointed in any direction, and that they are set at a slightly divergent angle with each other. By this arrangement of tubes—that now most approved by the naval authorities of Europe—two torpedoes are discharged at the same instant, and it is believed that the slight divergence of their courses will greatly increase the chances of either one or the other striking the enemy's ship.

An ingenious application of electricity is employed to explode the projectiles discharged by the pneumatic guns with which that remark-able craft, the United States dynamite cruiser *Vesuvius* is to be armed. These projectiles are in fact aërial torpedoes, and for the fifteen-inch guns now proposed for the Vesuvius, they will carry a charge of six hun-dred pounds of explosive gelatine. High explosives, such as dynamite, gun-cotton, and explosive gelatine, require to be detonated in order to develop their full energy; and for this purpose fulminate of mercury has been found to be best adapted. It has been found, also, that a very much greater destructive effect upon a target is obtained when the explosive charge of an elongated projectile, like those to be used in the guns of the *Vesuvius*, is first ignited *at the rear end*, or "base," instead of at the point, and recourse was had to electricity to accomplish this result. The fuse consists of a fine platinum wire—wrapped with a fibre of gun-cotton and embedded in fulminate of mercury—connected with the circuits of two small galvanic batteries carried within the projectile itself. With either of these batteries active, the platinum wire becomes instantly heated to redness and explodes the charge. In one, the electri-cal connection is kept broken until the projectile strikes some solid body, such as a ship or a target, when the shock of impact closes the circuit and causes the explosion. The circuit of the other battery is complete; but the battery itself requires to be moistened in order to become active, and the explosion consequently takes place after the projectile has sunk below the surface of the water.

The use of electricity for the propulsion of boats has met with considerable success. But the low speed hitherto attained by such vessels, and the fact that their motive power is necessarily derived from storage-batteries, which are notoriously uneconomical, have prevented their use as torpedo-boats. Yet, next to great speed, the most essential quality of a torpedo-boat is noiselessness, and this requisite electric boats possess in an unsurpassed degree. This latter may even become, under some circumstances, of more importance than high speed; for on a dark night, or in a fog, the success of a torpedo boat will often depend upon a silent, unperceived approach, and the suddenness of its attack. Hence, it appears safe to predict that electric torpedo-boats, armed with both auto-mobile and the ordinary "spar" torpedoes, will find a place in the navies of the future.

For *submarine boats* no motive power yet discovered seems comparable with electricity. High speed in vessels of this class will probably never be of the first importance, since they are mainly designed for attacking ships at anchor. The part such vessels are destined to play in naval warfare is an interesting and unsettled question, and their development is being closely watched by the naval powers of the world. Hitherto, when subjected to rigid practical tests, they may be said to have been unsuccessful.

Lieutenant Fiske, of the United States Navy, has

Deck-plan of Yarrow Torpedo-boat, showing discharging tubes mounted on turn-tables.

designed very recently an electrical "range-finder," for measuring the distance of an enemy's ship or of a target, but at the present writing the details of his invention have not been made public. This same officer was engaged for some time on board the steel cruiser *Atlanta*, in conducting a series of experiments with the object of devising some means of telephoning or telegraphing, either through the air or the water, in order that ships might be enabled to communicate with each other in a fog, where signals could not be seen, or at night, when, as in the proximity of an enemy, the use of signal-lights might not be advisable. But while "results were obtained which were in themselves encouraging, they seemed to indicate the impracticability of the scheme."*

In the French Mediterranean fleet, during recent naval manœuvres, a novel use of the telephone was made in connection with a balloon. The latter, carrying with it a telephone connected with the ship, was secured to the deck by a rope which allowed it to rise to the height of about a thousand feet. From this lofty point of observation an officer watched through a telescope the operations of a distant squadron representing the enemy, and from time to time telephoned the results of his examination to those on the vessel below.

A number of other applications of electricity, such as electrical steering-gear, electrical engine-room annunciators, etc., have been made on ships of war; but they have either proved to be unsuccessful, or, if successful, they were found to be of such doubtful utility as to render their adoption unadvisable.

The development of electricity is so rapidly progressing that no limit to its future naval applications can now be assigned. But, with the assistance of the Office of Naval Intelligence, it may be safely assumed that the proposed "New Navy" of the United States will have the benefit of every useful naval invention. This office, to which the writer is greatly indebted for information and illustrations included in this paper, is a branch of the Navy Department specially charged with the duty of keeping informed upon the latest advances in all branches of naval science throughout the world.

* Report of the Office of Naval Intelligence.

ELECTRICITY IN LAND WARFARE.

By JOHN MILLIS, U. S. A.

Invention Mitigating the Evils of War—Shorter Conflicts, but More Severe—Application of Electric Devices to Military Purposes—The Initial Velocity of a Projectile Determined by an Electric Chronograph—The Bore of Large Guns Examined by Incandescent Lamps—Photograph of a Flying Bullet by Electric Light—The Electrical Target—The Field Telegraph in Military Operations—The Telephone on Battle-fields—Uses of the Captive Balloon—Signalling by Search-lights—Electric Exploders for Guns and Mines—A Battery in the Shell of a Dynamite Gun—Submarine Mines—Torpedoes Electrically Steered from Shore.

IT is not an uncommon observation that all the great inventions and discoveries which have contributed to mark the present century as the most extraordinary in the history of the world's progress have been immediately applied, so far as possible, to the "abominable end of annihilating mankind;" but though it is true that a majority, perhaps, of the improved appliances of modern warfare have thus resulted from the application to military purposes of discoveries which first found practical use in the industrial or commercial world, the high degree of perfection which has been attained in the development of military methods and *matériel* of the present time is also largely due to the special devotion to this particular object of the best educated intelligence and inventive talent.

To the reader accustomed only to the peaceful occupations of business or professional life this may seem like a perverted application of human knowledge and technical science, and one entirely inconsistent with modern ideas concerning the proper relations which should exist among civilized nations. It is evident, however, on reflection, that there can be no possible insurance for the nation against attack and defeat by a foreign power, except the most perfect preparation to meet and resist such an attack that the skill and knowledge of man can devise. Moreover, from a humane point of view, it is a gratifying reflection, and a fact fully established by history, that with the improvements and advancements in military science and art of modern times, warfare now consists of much shorter conflicts in arms than was formerly the case, and though the action is thereby rendered vastly more intense while it continues, it also becomes the more decisive, hostilities are terminated the sooner, and the loss of life is actually less; so that instead of contributing to the " abominable end " of exterminating the human race, the application of the most advanced discoveries and inventions to the art of war has in reality mitigated the distressing results of armed conflict.

The numerous and constantly increasing adaptations to the needs of man of that wonderful and still mysterious force or agent called electricity have naturally attracted great attention in the military world, both from the inventor who is inspired by the hope of a large pecuniary reward, and also on the part of those who have adopted " the service " as a life work.

Since a great portion of the details of the so-called " operations of war " are in reality only the operations of peace greatly intensified and prosecuted under more difficult and exacting conditions, it will be seen that nearly all the industrial uses of electricity find a more or less direct application to military purposes ; but it is proposed to present here only a brief review of some of the more important of these applications, together with references to certain uses of electricity which are peculiar to the military art.

It must be borne in mind that the preparations for war which are made in time of peace are often of necessity more or less tentative and experimental in their nature, requiring the test of actual service in time of war, or, at least, in the field, to determine their practical utility, to develop their defects, and to suggest the remedies. It would therefore be futile to attempt a reference to many ingenious and promising devices for use in war which have been suggested and made possible by the more recent advances in electrical science, but for the actual trial of which there has, fortunately, been as yet no opportunity.

Beginning with the permanent military establishment in time of peace, we find that electricity renders important assistance in maintaining this establishment in a proper state of drill, discipline, and general efficiency, as well as in the manifold technical processes connected with the fabrication of arms, ammunition, equipments, and other war *matériel*.

The soldier must first be supplied with an efficient weapon, and in the design, manufacture, and test of improved fire-arms many intricate problems arise, the solution of which could not be satisfactorily effected without the aid of electricity. One of the most interesting of these problems relates to ballistics, or the scientific study of the motion of a projectile when fired from a gun. It is evident that in order to predict the range of the shot, and its penetration or effect on striking the object fired at, and to determine the accuracy and efficiency of the gun, of the powder, and, to a certain extent, of the projectile itself, an accurate knowledge of the velocity imparted to the shot when the weapon is discharged is of the greatest importance to the ordnance officer. The rate at which the projectile moves when leaving the muzzle of the gun is called its *initial velocity*, and since its rate of motion rapidly diminishes during flight, the observations to determine this initial velocity must be confined to a short portion of the projectile's path immediately after it is clear of the gun.

With modern high-power guns the shot is often given an initial velocity of more than two thousand feet per second, which is at the rate of a mile in less than three seconds, and hence it will be understood that the determination of the initial velocity involves, besides ascertaining the exact instant at which the shot passes two fixed points near the beginning of its path, the precise measurement of an extremely small interval of time.

Suppose a wire to be stretched across in front of the muzzle of the gun, and at a short distance from it, this wire forming part of a complete electric circuit, which includes also an electric battery and an electro-magnet. When the gun is fired this wire will be broken, the circuit interrupted, and the armature of the magnet released at the instant that the projectile reaches the wire in its flight. Now imagine a second wire stretched across the path of the shot, at a point one hundred feet in front of the first one, with a similar arrangement of electric circuit, battery, and electro-magnet. The breaking of this second wire by the flying projectile will release the armature of the second magnet, and if the interval of time between the release of the two armatures can be determined, the velocity of the projectile is easily calculated. If the projectile has a velocity of 2,000 feet per second, and the wires are 100 feet apart, the interval of time will be one-twentieth of a second, and since it is necessary to know the velocity of the shot to within a few feet per second, an instrument capable of measuring the time-interval to within the five-thousandth part of a second is required; and the instrument must also record its measurement, since the mind cannot conceive, and much less note, such a very brief period.

These exacting conditions are completely fulfilled by the electric chronograph, which is an instrument embodying the two electro-magnets already referred to, together with a recording apparatus, and certain attachments for making necessary adjustments, etc. The armature of the first magnet is an iron rod about three feet long, which is suspended in a vertical position from the core of the magnet. The breaking of the first wire releases this rod, allowing it to fall, and when the second wire is broken the armature of the second electro-magnet, which is placed a short distance below the first, is also released, operating a sort of knife which strikes the side of the falling rod, and makes a cut or indentation. The distance of this cut from the end of the rod shows exactly how far the rod dropped while the shot was passing over the distance between the first and second wires, and the corresponding interval of time is readily computed from known laws of falling bodies. Knowing this interval and the distance between the two wires, the initial velocity of the projectile is easily determined.

In testing large guns at the United States Proving Ground at Sandy

Measuring the Velocity of a Cannon-ball by Electricity.

Hook, New York Harbor, two open frames or targets are set up in front of the gun, and the wire to be broken is strung many times back and forth from one side of the frame to the other, to better insure the rupture of the circuit by the shot as it passes through the frame. From these targets conducting wires lead to the laboratory near by, where the chronographs, batteries, switch-board, and other apparatus are located. In order to obtain more reliable results, two or more chronographs may be used for testing the same shot, and the laboratory at Sandy Hook is provided with a number of these instruments.

Another interesting application of electricity at the Proving Grounds is the employment of a small incandescent electric lamp, mounted in connection with a magnifying mirror on the end of a long rod, for examining the bore of large guns. By the aid of this apparatus most thorough inspection can readily be made of the condition of the chamber and rifling, and any erosion or wear, which frequently occurs from the action of the projectile and the gases from the powder, can at once be detected. Photographs of the interior of the gun are even taken by means of the electric light, and the information obtained by these means is most valuable, and is manifestly more satisfactory than that given by the method of taking impressions of the bore in soft rubber.

Photography has also been made use of in studying the motion of the projectile. In the case of large guns, instantaneous views of the shot during its flight have been successfully taken by means of a camera provided with a quick-acting shutter. This method is not applicable, however, to small arms. A rifle bullet is a very small object, and the camera must be set very near its path in order to obtain a picture of sufficient size to be of use; but the nearer the instrument is placed to the moving object to be photographed, the more rapid is the motion of the image over the plate, and no "instantaneous" or quick-acting shutter could possibly be made to operate with sufficient rapidity, or at the proper instant, to give a sharply defined picture. The desired end is accomplished, however, by the aid of electricity. The camera is provided with an extremely sensitive plate and placed in a dark room, through which the bullet is made to pass. The instant the bullet is in front of the camera it breaks an electric circuit, producing a spark which illuminates the bullet for an instant, and its image is impressed upon the sensitive plate. The duration of the electric spark is almost infinitesimal, and since the plate is affected only during the continuance of the spark, a well-defined photograph of an object moving at a greater velocity than that of sound is obtained. Such pictures show the condensation of the air in front of the bullet, the vacuum left behind it, and the eddies and currents produced in the surrounding atmosphere by its motion; and they afford information which is of value in determining the best shape to be given to the projectile in order to reduce to a minimum the resistance which the air opposes to its flight, and so increase its range and effect.

The gun and ammunition having been brought to a proper degree of perfection for efficient service, electricity again comes to the soldier's aid in assisting to teach him how to use his weapon.

The electrical target, invented by an officer of our service, is designed to remove the dangers of small-arms practice and to facilitate instruction in marksmanship. This target is made of metal, its surface consisting of a great many separate plates accurately fitted together, and each one having a circuit-closer at the back, and admitting of a slight movement independent of the others. Each circuit-closer is connected by a wire with an annunciator located at the firing stand. When a shot strikes any part of the target the corresponding plate is slightly depressed, its circuit-closer operates, and the position of the shot on the target is at once indicated by the annunciator. The necessity of having a man at the target to mark the shots is thus obviated, and mistakes in marking, as well as accidents, which frequently happen from firing when the marker is exposed, are prevented.

As the telegraph was the pioneer among the great industrial applications of electricity, so it was also one of the first electrical inventions to be systematically and generally adopted as a necessary adjunct to the methods of warfare, and its recognized importance will be understood from the fact that the military organizations of nearly all civilized nations now comprise a special telegraph or signal corps, which is instructed and trained in all the duties and operations pertaining to the electric telegraph. Indeed, it has recently been remarked by a foreign military authority, that "so important will be the part played by military telegraphy . . . that the army which has the most efficient system of electric signalling will hold a trump card which may be a decisive influence on the game in any future European war."

Telegraph lines, constructed primarily for commercial purposes only,

15

render valuable service in the various duties connected with the regular maintenance of the permanent military force, and upon the approach of hostilities such lines may prove indispensable. The army must then be promptly mobilized, or placed in a condition for active service, and means of rapid communication between the capital or general headquarters and the centres of the various separate organizations is essential: as is also the prompt transmission of the news of the approach of the enemy, of his arrival on the sea-coast or frontier, and of his movements generally.

The army having been mobilized and placed in the field, the commanding general must maintain communication with the home government and with his subordinate commanders, who may be at a distance. In all these cases commercial telegraph lines are utilized to the fullest extent possible, as well as under any other circumstances where such lines exist and can be made available.

Permanent telegraph lines are frequently erected, especially for military purposes, to connect established posts, depots, or stations with each other and with headquarters, or to insure communication between other places which in time of war are likely to become of strategic importance. Such military telegraph lines being constructed in time of peace are, of course, similar in all respects to commercial lines.

But in active operations in the field, in presence of the enemy, and perhaps in the enemy's country, entirely new conditions are met with. The commanding general may find himself in a region where none of the conveniences of civilization ever existed, or where all such conveniences that may have existed have been destroyed. His command may extend over a large area, and some of the corps or divisions may be miles away, with dense forests, marshes, streams, mountains, or country occupied by the enemy intervening. Communication must be established through the nearest point to which the permanent lines are still in operation with the home capital, and it is absolutely essential that the commander be provided with the means of transmitting his orders to his subordinates, and of receiving intelligence of their posi-

tions and movements, and of those of the enemy, if he is to have any
hope of successfully executing the plan of operations which it is his
duty to carry out.

The Field Telegraph Corps now comes to his aid. This corps is
provided with its train of wagons or carts, drawn by horses or mules,
and carrying wire, poles, instruments, etc., and the tools and material
needed in erecting the lines, in taking them down, and in making neces-

Running the Wires of a Field Telegraph.

sary repairs. For lines which are to be in a measure permanent, like
those erected to maintain communication between the field of operations and the home government or the general headquarters, the wires
will be put up in the ordinary manner, making use of such available
material for poles, etc., as may be found in the vicinity.

In the case of temporary lines, which must be more hastily constructed and removed, the wagon carrying the wire is driven over the

route and the wire reeled out along the ground, to be afterward raised and supported on insulators attached to trees, buildings, fences, etc.; or, where such supports are not available, to light wooden poles carried for the purpose. The batteries, instruments, etc., can be set up in the wagon, which is thus made to serve as an operating station.

For still more rapid work, wire or cable having an insulating covering throughout is used, and this is reeled out along the ground, crossing marshes, streams, etc., if necessary. It is then ready for immediate use without requiring insulating supports. In crossing roads, and at other points where the wire would otherwise be exposed to injury, it is raised on poles or may be buried in the ground. In rough, mountainous, or thickly wooded regions, where wheeled vehicles cannot be used, the wire and other material is transported on pack animals, and in running short lines which must be very hastily established in difficult situations, the reels may even be carried by men on foot.

For operating the line an instrument similar to the ordinary Morse apparatus, arranged in compact form, is generally used; but the telephone has also a limited application for the purpose, and an improved instrument has recently been devised for the field telegraph service which will work for moderate distances through an ordinary bare wire laid along the ground or through the wire of fences.

In the use of captive balloons for observing the strength and position of the enemy or of his works, or for obtaining information regarding his movements, the telegraph or telephone furnishes the best means of communication between the observers in the car of the balloon and persons on the ground. The instrument in the balloon is readily connected with that below by wires carried along or enclosed within the cable by which the balloon is secured, and the results of the observations, instructions or orders, etc., can be communicated almost as easily as between two stations on the ground.

The members of the telegraph corps must, of course, be thoroughly drilled in all the manoeuvres connected with packing, transporting, erecting, operating, and taking down the special apparatus of the field signal train, and they must be able to operate the apparatus in commercial use as well. Besides having a thorough knowledge of the ordinary or Morse code, they must be familiar with various secret codes and ciphers, adapted for use in presence of the enemy, and they must be

Telegraphing Military Observations from a Captive Balloon ; the wire along the cable.

instructed in the methods of destroying or interrupting the telegraph lines of the enemy and intercepting his messages.

Recent improvements in synchronous telegraph apparatus, the action of which depends upon the continuous motion in perfect unison of two instruments in electrical connection, but at different stations, have rendered it practicable to make, at a distant station, an exact reproduction of any writing or drawing executed on suitably prepared paper and placed in the instrument at the sending station.* Though the principle of this instrument is not new, and it has never been used to any extent for commercial purposes, the improvements referred to give promise of rendering the system of great value in sending vastly more complete and reliable information concerning the plans of forts and other works of defence, the disposition of troops, maps of the country, etc., than could possibly be conveyed in words, and with a great saving in time and almost absolute security against the interception of the message by the enemy.

But even the field telegraph may fail, as when it becomes necessary to transmit messages over large bodies of water, or across intervening country, which, on account of its occupation by the enemy or from some other cause, is inaccessible. Flashes of light, produced by the field search-light apparatus, directed upon the clouds or upon high terrestrial objects in the vicinity, may then be employed, or signals may be sent by means of incandescent lamps sent up in a balloon and made to flash by manipulating a key introduced into the circuits. This is a method of signalling which cannot be interrupted by the enemy, though it is, of course, only available at night and in clear weather.

The search-light apparatus consists of the well-known arrangement of a powerful electric lamp placed in the focus of a reflector, so contrived that the lamp and reflector can be easily turned and the beam of light sent in any desired direction. For field service, the lamp and the steam-boiler, engine, and dynamo for operating it are mounted on wheels for facility in transportation. The more important uses of this light are to disclose the position and movements of the enemy; to blind and confuse him in the event of an attack at night; to enable the fire of a fort to be properly directed; to facilitate night movements of the army, and the construction of earthworks or other hasty defences. After an action the light may also be employed to search out and care

* See "The Telegraph of To-day."

Using the Search-light on a Battle-field—Bringing in the Wounded.

for the wounded on the field of battle, and it assists in the merciful work of the surgeons by enabling them to promptly perform delicate operations which otherwise could only be done in daytime.

It is, however, in immediate connection with the high explosives that the military adaptations of electricity have received their most perfect development and have contributed to the most astonishing results, and there is probably no military weapon with which the gen. eral public has been made more familiar by repeated descriptions and illustrations than that class of instruments of destruction known under the general name of torpedoes.

The almost universal adaptability of the method of firing charges of gunpowder, dynamite, gun-cotton, or other explosives, by means of the electric spark or an electric fuse, follows from the great ease and facility with which the firing circuit can be established in almost any situation, however difficult or unfavorable, and carried to almost any point, however remote; thus enabling the operator, by simply manipulating a key or electric exploder, to fire the charge from a place of perfect security, with certainty and at the proper instant. There is no "slow match" or powder-train liable to be ruptured, to become damp, or to be found inoperative from other causes when most needed; and the unsatisfactory percussion cap and lock, always subject to accidental discharge, are also done away with.

More especially in permanent fortifications, and in the complete and complicated systems of defence established to protect important seaports, does the electric apparatus, or "plant," become an important factor in the organization, and under these circumstances it requires a special corps of skilled operatives carefully trained in its establishment, care, and manipulation.

The main station, containing the steam-power, dynamos, batteries, etc., is located in a thoroughly secure place within the work, or it may even be removed to a point some distance away and the conductors carried to the fort under ground. Electric motors will enable the aiming and manipulation of the heaviest guns, mounted on "disappearing gun-carriages," to be effected with the greatest facility, and they will also be brought into use in moving and hoisting the enormous charges of explosive and projectiles which constitute the ammunition of modern high-power ordnance.

The guns have electric fuses, or primers, connected with wires which are led to a secure place of observation, where the commanding officer, provided with instruments which enable him to accurately determine the position and distance of the enemy's vessels, has the entire battery of the fort under immediate control, and he can direct and regulate the fire of the guns to the best possible advantage.

Torpedo Explosion under Ice.

The approaches to the fort by land will be defended by ground torpedoes or mines, properly distributed and carefully concealed, and passages over frozen streams or other small bodies of water in the vicinity may be defended in a similar manner. Mines placed under ground or under the ice are also employed in field operations, and frozen bodies of water are readily made passable for friendly boats by blasting with charges of explosive placed on the ice or beneath the surface.

These mines or torpedoes constitute a formidable and much-dreaded means of defence, since they are entirely concealed from sight and their position cannot easily be discovered by the enemy. By providing them with electrical fuses, placed in circuits under control from the fort, they will be absolutely inoperative and harmless until, upon the approach of the enemy, the time arrives for their deadly work. The mines can also

A Mine Explosion during an Advance.

be arranged to operate automatically if liable to be disturbed during darkness, or if planted where their location cannot be observed from the fort.

The electric light is also a valuable aid in the general system of defence. The incandescent lamp is particularly adapted to illuminating dark passages and the magazines in which the ammunition is stored, where, on account of the danger of fire, any other form of artificial light has to be employed with the utmost caution.

In the pneumatic dynamite gun, which has recently received so much attention both from military men and from the world at large, and which undoubtedly will have a place in shore defence, electricity is employed to regulate the explosion of the charge carried by the projectile. The shell carries a small battery in one of its compartments, and this battery is entirely inoperative under ordinary circumstances, rendering the shell perfectly safe to handle and place in the gun. When the gun is fired, should the shot fall short, the battery becomes active soon after immersion in the water, and the explosion consequently takes place below the water-line, where it is most likely to do injury to the vessel aimed at.

But it is in the system of defensive torpedoes, or submarine mines, which is provided for obstructing the entrance to the harbor, and which is operated from stations on shore, that we find perhaps a more elaborate and complete arrangement of electric apparatus than in any other portion of the defensive establishment.

The conditions imposed upon this part of the defence are many and exacting, and a vast amount of study and experiment has been devoted to its development and perfection.

A submarine mine consists essentially of a water-tight metallic case containing a large charge of dynamite or other powerful explosive and provided with an electric fuse and a circuit regulator of peculiar construction. An anchor or sinker is placed on the bottom of the channel it is desired to obstruct, and to this the mine is moored so as to float a short distance below the surface. The fuse and circuit regulator are connected by means of a cable with the testing and firing apparatus in the operating room, which is located in a secure place within the work on shore, and a large number of such mines are distributed over that portion of the channel through which an enemy's vessel must pass in order to enter the harbor.

Those mines which are placed in the main channel or entrance must be under most perfect control of the operator, since they are planted before the arrival of the enemy, and the channel must afford a safe passage for friendly vessels **until** hostile ships attempt an entrance. With this object the apparatus can be adjusted so as to merely give warning **if** the torpedo be struck or disturbed **in any way,** leaving the matter of firing entirely to the will **of the operator or to that of the officer in com-**mand of the defence, who is stationed **in a suitable place for observation** and in telegraphic **communication with the torpedo operating-room.**

By making suitable connections **with the shore firing** circuits, guns ready loaded and aimed **over the places where the** mines are planted are automatically discharged as soon as a mine is disturbed, thus preventing boat parties **from** tampering with the **system,** while the mine itself is reserved for **its proper work.** Finally, **the** mine and gun circuits may be so arranged that either **the** mine alone, **or** both mine **and guns are** automatically discharged at the proper instant upon the attempted passage of a man-of-war.

There are, besides, a great **many other** details connected with this system which cannot be referred to here, but which **are,** nevertheless, important, since the efficiency of the entire system depends largely **upon** careful attention to these very details. **For example, it is** necessary that provision **be** made for accurate electrical tests **and measurements,** both while the mines are being planted **and after they are established in order** to determine with certainty their **condition of efficiency** at all times, and to make the best use of such as may give signs of deterioration ; and it will be readily understood that with such large charges of deadly explosive and apparatus which is necessarily somewhat delicate and complicated, the greatest care must be exercised in **order to prevent serious** accidents.

Movable or fish torpedoes also form part of the **shore-defence** system, and in one of the forms of **this** torpedo adapted to this purpose the weapon is propelled through the water **at great speed by** an electric motor connected to the propeller-wheel and driven by **an** electric current conveyed **to it from the** dynamo on shore **through a cable which** the torpedo pays out while **in** motion. The steering apparatus is also operated electrically from shore through a separate conductor contained in the same cable, and the charge of dynamite which the torpedo carries is exploded either at will or automatically upon striking the enemy's vessel.

Although extraordinary results may reasonably be expected from the employment of the electrical torpedo and dynamite gun for coast defence in future wars that may occur, it must be admitted that there is a popu-

The Sims-Edison Electric Torpedo, steered from the shore.

lar tendency to overestimate their value and the security which their use will afford the country. Granting that they both are now recognized weapons of warfare, it must be remembered that neither one alone, nor both together, are able to cope with an adversary fully equipped with both these and with other modern weapons besides. To provide only torpedoes and dynamite guns for the country's land defence would be as great folly as to send an army into the field with only the modern artillery arm, knowing that it would have to encounter an enemy armed with the same weapon and with the rifle and sabre as well.

ELECTRICITY IN THE HOUSEHOLD.

By A. E. KENNELLY.

IT would be strange, indeed, if so readily controlled an agent as electricity, an Ariel before whom time and space seem to vanish, did not cross the threshold of our homes and enter into our household life. We find, in fact, that the adoption of electrical household appliances is daily becoming more widespread, here adding a utility, and there an ornament, until in the near future we may anticipate a period when its presence in the homestead will be indispensable.

The first application of electricity to household purposes was presented by the electric bell early in the century, and annunciators of various kinds soon followed. For many years this was the only convenience it afforded; but the discoveries of the telephone, the electric light, and the electric transmission of power within the last thirteen years have given it a tremendous impetus whose ultimate consequences are not yet within view. Even if, as seems unlikely, these brilliant

achievements are destined to stand alone, not succeeded by further discoveries, many years must elapse before their full use shall have been
reached; just as in the case of the pianoforte, which took more than a
hundred years from its first invention to become the common guest we
find it in the household of to-day.

In the electric bell, the pressure of the finger on a button brings two
strips of metal into contact and completes a circuit, forming, as it were,
an electrical endless chain from the battery through the wires, bell, and
annunciator. The whole circuit instantly gives passage to a current of
electricity, and in consequence becomes endowed with magnetic properties throughout. By means of an accumulation of wire, as a coil round
a horseshoe bar of iron, the magnetism is locally intensified to an extent necessary for the attraction of the iron hammer bar, and by a
simple automatic device the blow on the bell is reduplicated. A similar
electro-magnet in the annunciator releases by its pull a shutter, indicat-

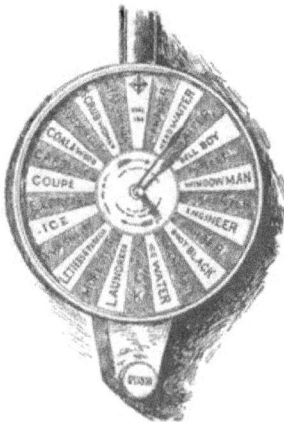

An Electrical Call.

ing the room whence the call has come.
No system can be imagined more simple,
and in spite of many an overtasked battery or dust-invaded indicator, it every.
where holds its own. To put mechanical pull-bells into a modern dwelling is
an anachronism.

The same principle is the basis of
every annunciator system, with such
modifications as improvement in the
particular direction of the design may
have suggested. Even those complex-
looking annunciators to be met with in
large hotels, which by means of a dial
in every chamber enable its inmate to
call for almost any common requirement, from a newspaper to a
complicated beverage. differ from the general plan only in their
power to signify a particular summons by the aid of a definite

number of successive contacts and corresponding electro-mechanical impulses. A good example is afforded by the burglar-alarm apparatus. Every door and window through which entrance could be forced is fitted with a simple clip, adjusted to make, on the least opening, a metallic contact which sets an alarm bell in operation, and at the same time indicates the room where the invasion is being made. By means of a small key, or "switch," the battery is cut off during the day. Such a system adds greatly to the security of a household, and only needs occasional regular supervision, since all the contacts are necessarily somewhat exposed to dust and moisture. A trial once a week is a matter of a few minutes only, and is amply repaid by the greater sense of security it gives. It has been said that a burglar would soon ascertain whether a house were so guarded, and that before opening a window he could, by removing a pane, find means to cut the electrical wire connection at the sash. This objection is, however, invalid, for the system can be easily arranged to give the alarm equally well for any disconnection so made.

Another most useful system, on the same plan, controls the automatic regulation of temperature. How much discomfort and indisposition would be saved in many a household if the temperature were constantly maintained in every apartment at the desired point, both in summer and winter, independent of irregularities of the season! So far as concerns our winters, this is quite within practicable limits, while in summer the temperature can always be moderated, if not actually kept uniform, by utilizing the controlling power of electricity. Thus in winter time, whether a house be warmed by water, hot air, or steam, it is only necessary to place in each room an automatic thermometer which makes a contact as soon as the temperature reaches the desired point, and to arrange that the contact so made shall electro-magnetically cut off the supply of heat from that chamber. The subsequent cooling of the room below the limiting temperature causes the thermometer to break the circuit and readmit the heat, and it is only necessary to keep an abundant supply in reserve in order to obtain a practically equable

16

temperature. Such a thermometer, generally called a thermostat, is made by riveting side by side two strips of different materials—generally brass and rubber—which expand differently at the same degree of heat. The composite strip so formed is warped by changes of temperature, which unequally affect the lengths of the components, and being free at one extremity while firmly fixed at the other, the effect of this warping is magnified into an appreciable range of movement at the free end. This enables a contact to be made at any point within that range, while a screw adjustment and dial arrange for the contact to take place at any temperature within desirable limits. The parlor thermostat can therefore be set at 70° while that in the hall is fixed for 60°. It is generally claimed by those who have adopted the system that a decided saving in fuel is effected, in addition to the comfort gained through the absolute prevention of overheating in any part of the house. The thermostats are so sensitive as to respond to the change of a single degree in temperature. The maintenance of the equilibrium, then, depends on the supply of heat and the facility for its distribution through each room when once admitted.

In the same way, during the summer months, this thermostat can, by an additional contact, control the supply of fresh or, if possible, ice-cooled air, so as to maintain a pleasant temperature within doors. Such a system has for two years been in successful operation at a large country house near Greenwich, Conn. In winter time it is warmed by fresh air drawn in through an underground pipe, and heated by passing through a reservoir in which a long steam-pipe circulates. Thence it is fanned into the different rooms through dampers, each controlled electro-magnetically by a separate thermostat. In summer the water-supply of the house, as it comes from deep wells, takes the place of the steam in the circulating pipe of the reservoir, and so cools the incoming air; the same thermostats adjusting the distribution. In this way the temperature is maintained throughout the house at 70° in winter, and does not exceed 75° in summer; while the ventilation is controlled by the same apparatus.

The fire-alarm system depends upon a similar thermostat set for higher temperatures, usually from 120° to 160°. The contact in this case rings an alarm bell and indicates the room where there is danger. It is hardly possible to overestimate the utility of a well-arranged fire-alarm household-system, which makes it possible to extinguish a fire in its beginning. Statistics certainly show a marked decrease, by the use of electrical fire-alarm systems, upon the number of serious fires in towns; but the conflagrations that have been saved by the timely local warning of domestic apparatus, report can never tell.

In some town-houses fire is not the only rebellious element over which constant watch has to be maintained, water overflow from tanks and bursting pipes being almost as much to be dreaded. The *Journal of the Franklin Institute* has called attention to an electrical device which is set in operation by a float, the contact so established cutting off the water-supply or indicating the danger as soon as a definite level is reached. An electric door-opener has also been lately designed by which visitors can be admitted without delay. The closing of the door compresses a powerful spiral spring, which is then held in check by a lever until the latter is released by an electro-magnetic impulse. The spring forces open the door, the latch at the same moment being withdrawn.

Of inferior importance to these systems, which guard the safety of the household, but yet of great interest and utility, is the clock system. Appreciation of time and its value is said to be the test of a nation's activity, and it is surely a luxury to see all the clocks in a house keeping an even pace. There are several methods in use for this purpose, and they form two distinct classes, one adopting centralized government, the other local administration. In the former a single clock as standard drives all the others electro-magnetically, their operation depending entirely on the electricity supplied during its periodic contacts. In the latter, each clock is a free and independent timekeeper, whose rate, however, is under regular electrical control from the standard. This control may be exerted continuously on the pendulums, but perhaps the simplest and most satisfactory household system yet tried is that in which the control is effected once in each hour. Exactly at the hour the standard clock makes a contact, completing a circuit through all the controlled timepieces, and electrically exciting a magnet in each. In obedience to this impulse, a pair of arms spring from the dial at the figure XII, and meet swiftly in the centre with the minute hand tight in their embrace, and vanish the next instant behind the dial, where they await the next hourly summons. Each clock is thus mechanically corrected every hour, as the arms sweep over three minutes' space on each side of the true vertical, and the clock that fails to keep time by three minutes in the hour may well be submitted to internal examination.

Another convenience which is sometimes added to a system of time regulation is an arrangement for electrically winding up the clocks at regular intervals. So long as the electrical supply is maintained, and the clock-work continues in proper working order, such a system forms as near an approach to perpetual motion as the conditions of our planet give us the right to expect.

The electrical time-detector is an instrument much used in large buildings over which continual supervision is needed. It serves to register the time at which visits are paid to any particular part of the premises, and, in fact, successfully solves the problem of keeping watch upon the custodian. A dial, rotating by clock-work once in twelve hours, carries round a paper disk over a perforated metal plate. Each push-button in the house controls, by its own pair of wires, one electro-magnet, the armature of which, on attraction, punches a hole in the paper disk through a particular aperture in the plate. This hole is always in a certain ring marked for the purpose. The watchman going round the building pushes the various buttons on his way, thus registering his progress on the paper disk by punched holes: the rings marking the buttons and the angular position indicating the time.

The discovery and introduction of the electric telephone has marked an era in the annals of household affairs, as the existence of four hundred thousand in the United States to-day amply attests. The economy of time its use has effected is incalculable. /Its greatest fault is perhaps an occasional tendency to mingle the speech of one interlocutor with the conversation of less interested neighbors. Within the limits of a residence, no better interior communication can generally be had than by the ordinary speaking-tube; but in connection with outbuildings on an estate, the telephone is a great advantage. When several such houses are connected by telephone with the main building, it is possible to arrange that any two can communicate with each other on the same wire without calling the attention of the rest, a system saving much time and trouble. As many as eight telephones are sometimes worked in this way on the same wire, and although only two can employ the line at one time, the calling of any particular person is not heard by the others.

Among the greatest gifts that electricity has bestowed on domestic life is the incandescent electric light. There can be little doubt that, when experience shall have given confidence in its trustworthiness, while time shall have rendered its many excellences familiar, it will be adopted in all households. It neither consumes nor pollutes the air in which it shines, whereas the ordinary sixteen candle-power gas-burner vitiates the atmosphere with its products of combustion to the same extent as the respiration of five persons. Besides, those products ultimately injure books, paintings, and ceilings continually exposed to their influence. As the gas-jet develops some fifteen times as much heat as the electric lamp of equivalent power, the latter adds greatly to the comfort of a house in warm weather. In the nursery it is particularly welcome, for it requires no matches, cannot set fire to anything, even if deliberately broken while lit, and effectually checks the youthful tendency to experiment with fire.

In addition to this, its complete amenability to control, and submission to all change of position or equilibrium, render it everywhere admirably adapted to the purposes of adornment. Some of the most charming effects can be produced by good taste in the choice of centres of illumination, together with appropriate surroundings.   In the parlor an illuminated painted vase, lighted from within, may vie in attractiveness with the pictures on the walls, whose colors are almost as readily appreciated by incandescent as by day light, while opalescent globes of varied shade tone the brightness everywhere into subdued harmony.   In the billiard room the table is brilliantly lit, without danger of soot or oil marring the baize, and on the veranda the lamps shine heedless of the wind.   A very pretty effect can be also produced in conservatories, by suspended lamps of different colors half hidden in the foliage.

The electric light can also be made to give a very beautiful effect in illuminating garden fountains.   For this purpose a chamber has to be excavated beneath them, and immediately under the jet a thick plate of glass is inserted, water-tight.   An arc-lamp directs its light directly through this plate into the column of water rising vertically above it, and the enclosed air, together with the broken surfaces of the jet, scatters this light in all directions, thus giving the liquid the appearance of being self-luminous.   The color of the illumination is varied by means of tinted slides passed horizontally beneath the glass plate in the roof of the vault. A very handsome display of this description was made at the Paris Exhibition this year.

The steps in the development of an incandescent lamp during manufacture have been traced in the chapter on "Electricity in Lighting." When the completed lamp is placed in circuit, the carbon filament conducts electricity but only imperfectly, and the latter thus requires a certain pressure to force it through the lamp.   The work done in overcoming the resistance so offered is developed into heat proportional to the square of the velocity of flow.   The frictional opposition of a pipe to the passage of water it conveys, generates heat at greater rate than the square of the velocity ; but these two cases of motion present many analogies, although the pipe deals with the transmission of matter itself, while the filament deals with the transmission of a condition of matter only.

At a certain electrical pressure on the filament the right quantity of electricity flows through it to bring its temperature to the incandescent point of due candle power. At this pressure the lamp will last probably two thousand five hundred working hours. If our best microscopes had a magnifying power **perhaps ten thousand** times greater than that they now reach, and it were possible to subject the glowing **filament** to their examination, we might expect to find the ultimate particles or molecules of carbon vibrating and colliding with **an** intensity that now baffles the imagination. **We can fancy that at the surface of** the filament **an** occasional molecule, projected outward with more than usual force, would bound beyond the range of retractive **influence, and** be hurled past **recall** (like the celebrated projectile of M. **Jules Verne) against the** distant inner surface **of the glass** globe. Gradually the latter would be darkened **by** the thickening meteoric accumulation, while **the** filament would weaken, as its dwindling substance (enduring such tremendous internal commotion) suffered structural decay, until **at some** point disruption would ensue, followed immediately by loss of **conduction and** extinction of the light.

The greater the **electrical** pressure **brought to bear upon the** lamp, the higher the incandescence attained. The lifetime of a lamp, endowed at the **outset** with average vitality, thus depends entirely on its treatment, and **can be** made almost what we please, from a few moments to even many years, according to the degree of incandescence it is called upon to produce.

In fitting up a house with the electric light, **a** little consideration is required to obtain the greatest convenience. The switches by which the lamps are turned on **and off** should usually be placed just inside the door, where they can be reached on entering or leaving the **apartment.** In the bedrooms, however, they should be suspended from the ceiling in such a manner **as to** be accessible **on** first entry, **over** a bracket by the door, and then movable to within easy **reach** of the bedside; or, better still, there may be two alternative switches—one at the door and the other by **the bedstead.** One test of **a** well-designed installation is that the householder should be able to visit the entire building, commencing with the hall door, from attic to cellar and back, without once being left in the dark, or leaving lamps burning on any floor behind him as he makes the journey.   A good plan, that has been carried out in more than one instance, is to have a spare lamp in each room under sole and direct control of the burglar and fire-alarm systems, in such a way that the forcing of any window, or any dangerous excess of temperature, may not only ring the alarm, but also light up the whole house.

In many cases where electricity is not itself the illuminant, the electric spark is often adopted for the purpose of lighting the gas.   In theatres, for example, a frictional electrical machine is employed which, when rotated by hand, is connected in succession to the various wires leading to different jets or clusters, and the sparks, passing between two metallic points set close to the burner, ignite the gas.   Similar arrangements on a smaller scale are in household use.   The pull on a pendent chain or the pressure on a button allows the current to pass from a battery through a small induction coil, the spark of which flashes at the burner.

The most ingenious apparatus of all, however, is the hand gas-igniter which, without any battery, produces a spark between two points in its tip on the simple pressure of a button on its side.   This compact instrument is, in fact, an electrical rotating influence machine (acting on similar principles to some of the most powerful generators of high-tension electricity), and it is difficult to realize that this safe and simple apparatus can produce sufficient electricity to light the gas, when the electrical pressure between the points at the moment of emitting a spark must be many times greater than that exerted upon an incandescent lamp.   Its operation depends upon the rotation of an internal cylinder which causes the initial charge to be augmented at a rapid and increasing rate until the tension is sufficient to create a spark between the opposed points.

The transmission of power is another application of electricity which has practically been evolved only within the **last decade, and** which is still in its infancy. Its usefulness in the household **is second** only to that of illumination. Ignorant as we still are of the real nature of this marvellous agent, we know at least that electricity implies power; all the evidences by which we are rendered sensible of its **presence are** manifestations of energy.

The electric motor is the **machine by which electrical power is ren-** dered mechanically **available.** Its principle is entirely magnetic; the pull that a wire conveying **an electric current is seen to exert** upon a compass needle in its vicinity **being** here enormously intensified by having **a large horseshoe electro-magnet for the compass needle,** and many turns of wire **close up within its grasp instead** of the single conductor. The revolving cylinder of separated copper segments on which the brushes **rest,** called the commutator, **is nothing** more than an electric treadmill, by which the current is cut off each wire in **turn as it** reaches the point of most powerful **attraction, so that the current is** always kept advancing toward the magnetic pole, **but never reaches it.**

The qualifications which peculiarly fit the electric motor for household use are its compactness, **perfect control, silence, and cleanliness.** It is a wonderfully compact **piece of mechanism, for in domestic sizes it** weighs under one hundred pounds per horse-power, **and its amenability** to control is evident from the fact that the turning of a switch will stop or start it. One great secret of **this compactness lies in** the fact that the motion is rotary, **and not oscillatory** like that of a piston; hence the great speed it can attain, **as also the absence of jar and noise in its** work. A small motor may thus become an **ornament, as well as a useful in-** strument. The illustration, on page 250, shows a Diehl motor attached to a sewing-machine spindle. In any house supplied **with the electric** light it **is only necessary to connect** the motor with the electric mains, like a lamp, and turning the **switch sets the machine at work, thereby** saving the hundredth part **of a horse-power,** which is the usual amount of energy **needed to drive it** by treadle, **not to mention the** comfort gained and nerve-force conserved.

Sewing-machine Run by Electricity; the entire motor concealed in the wheel-case at the left.

As another example of use and ornament united, circular fans driven by motors are not uncommon, and are luxuries in hot weather, when even the exertion of waving a fan counteracts the comfort so produced.

The electric motor is destined to enter largely into the operation of

The Electrical Fan.

elevators in town-houses, all its good qualities being in this case shown to advantage. In dwellings supplied with the electric light it is only necessary to fix in position a motor fitted with the requisite gearing, and connect the same to the elevator with wire ropes, the power being taken direct from the electric mains. In this respect, also, electricity, as a power-distributer, contrasts favorably with other sources in the reach of modern engineering. For, if elevators were to be operated from a central station by hydraulic power supplied to each house through pipes, then an elevator in motion would take as much energy from the station when empty as when fully occupied by passengers— unless, indeed, complicated devices were introduced to avert this waste. The electric motor, on the contrary, would, if properly selected, only draw from the mains the proportional amount of power required for the load to which it was subjected, in addition to what little it expended in overcoming the friction of its own mechanism, and consequently, so far as the supply of power was concerned, would be much more economical.

Another suitable task for the electric motor in country-houses is pumping. Where water has to be elevated from wells or cisterns to the attic level for household distribution, art and science lend the means, while electricity supplies the power. By the use of the rotary pump, the plant, which may be placed in the cellar, can be made wonderfully compact and quiet in its performance. How vivid is the contrast between this simple apparatus and the blindfolded horse, that, for the same purpose, has so often been condemned to describe endless circles, with a long trail-beam as radius and a well as centre. A float in the reservoir above breaks a contact as soon as the level of water there has reached the desired limit, and so automatically stops the motor until further supply is demanded.

In the same way motors have been applied to lawn-mowers, to carpet-sweepers, to shoe-polishers; and, in fact, there is no household operation capable of being mechanically performed, of which, through the motor, electricity cannot become the drudge and willing slave. It has even been applied to serving at table. A miniature railroad track runs round the table within easy reach of each guest, and thence, by ornamented trestlework, to the wall, disappearing through a shutter. The dishes, as electrically signalled for by the hostess, are laid on little

trucks fitted with tiny motors, and are started out from the pantry to
the dinner-table. They stop automatically before each guest, who,
after assisting himself, presses a button at his side and so gives the car
the impetus and right of way to his next neighbor. The whole journey having been performed, the cars return silently to their point of
departure.

The electric motor is also perhaps the most nearly perfect means
known of obtaining steady, smooth, and continuous mechanical motion,
and largely, with this object in view, it has been introduced into the
Edison phonograph, an instrument destined to play the very important
parts of music preserver, recorder, and amanuensis in the household of
the future. On the surface of its cylinder the delicate wavelets that
the voice has impressed sometimes cannot exceed the fifteen-thousandth
part of an inch, and on their due representation in vibrations of the air
the reproduction of the stored-up sound has to depend. The electric
motor enables all these to be reproduced in a manner that would not be
possible if there was any unsteadiness or tremor in the movement of
the working parts.

The motor also supplies parlor organs with air, and has been applied
to automatic pianos. A bright prospect also opens for the application
of electricity in country-houses, in the direction of artificial horticulture. Among the conditions that differentiate vegetable and animal
life there seems to be this remarkable fact, that plants do not essentially require sleep or periods of intermittence in growth and activity.
This is evidenced by the continuous and rapid growth of plants in the
far north during that brief but happy summer in which the sun never
sets. The electric arc lamp has been found, by the late Sir William
Siemens and others, to practically replace the sun in its effects on plant
life, over a somewhat contracted range, so that an extensive conserva-
tory lit by powerful arc lamps would be efficiently supplied for night
growth by some two candle-power per square foot of area. A hot-
house in reality artificially represents latitude in all respects save sun-
light, which the electric light is ready, in part at least, to replace.

Public attention has latterly been drawn to the question of electric heat-supply to houses, and it has been frequently supposed that the apparent novelty of the plan favored its commercial success. The fact is, however, that of all the practical applications of electricity there are none whose limits and possibilities are more clearly defined and better understood than heat distribution, for the simple reason that it has been attentively studied for the last ten years. This is apparent from the fact that the problem and task of electric lighting is, primarily and essentially, electric heating. Almost all the energy supplied electrically for the purposes of illumination is dispensed in the form of heat, and this heat is expended with the maximum economy that the engineering of the day permits in maintaining our carbon filaments at incandescent temperature. Despite the high economy in the consumption of power that the electric lamp possesses in comparison with combustible sources of illumination, it has lately been shown, by experiments at Cornell University, that only some five per cent. of this heat is yielded in rays of light, the remainder (at present essential to securing this result) being spent in raising the temperature of the air and surrounding objects. Consequently, whatever improvements the art of electric lighting may effect in economizing this large heat expenditure, and raising it into visible radiance, science appears to have determined that a given supply of electrical power can only yield the same amount of heat that it now develops in passing through our lamps. One form of electric heater operating within narrow limits might bring a piece of metal to melting-point, while another only slightly raised the temperature of a large volume of water; still, the total quantity of heat developed in each for a given supply of electrical energy would be precisely the same. The only economy that can be looked for in the distribution of heat lies, therefore, in saving the waste incurred by forcing electricity through the mains, and this is a margin that modern engineering has already rendered comparatively narrow.

Heat, being already distributed electrically on a large scale to houses for the operation of incandescent lamps, can be, and already has been,

applied for heating purposes exclusively. The difficulty of carrying
out this plan on a large scale, in order to replace household stoves and
furnaces, is a purely economical one. The question ultimately reached
is, whether labor can be saved to a community if all the coal necessary
for their heat-supply through the medium of electricity be burned in
one central station, and the electrical power so obtained distributed
generally, instead of continuing the usual custom of burning the coal in
each house locally. On the one hand, the local process of combustion
is at present a wasteful as well as a dirty one, most of the heat escaping by the chimneys; while, on the other hand, the steam-engine is
necessary in the central station to convert the furnace heat into electricity, and the best modern engines are only capable of utilizing twelve
per cent. of the heat developed from the combustion of coal under their
boilers; so that, when the machinery and conducting system of mains
are taken into account, the verdict (notwithstanding household smoke
and waste) has been hitherto against the economic possibility of the
electrical distribution of heat on a large scale. But every improvement
effected in the machinery for the conversion of furnace heat into electricity, every advance made in the progress of electrical engineering,
modifies in proportion the balance of advantages in this great social
problem, and it is well within the reach of possibility that electric heating may be as successful at some future date as electric lighting itself.
Even now there are many occasions where heat is required to be applied
very locally, in culinary purposes, for instance, and where the cleanliness and convenience of the electrical method might outweigh the objection of slight extra expense. The advantage to a man whose duties
call him out during the night of being able, from his bedroom, to set
an electric coffee-heater at work in his dining-room, so that by the time
he is ready to leave the house he finds hot coffee awaiting him, and all
without arousing any person in the house, far outweighs the three or
four cents for electrical power that the beverage has probably cost him.
Similarly, there are times when a foot-warmer is worth many times over
the expense of electrically preparing it at a few minutes' notice.

Both of these commodities are in actual use. / The stove is an ornamental case enclosing a coffee-pot, or, in another form, it may be a kettle in an asbestos lining, round which circulate coils of wire, the passage of the electric current through these coils generating the heat. In one convenient form the current that would feed fifteen ordinary incandescent lamps will produce hot coffee in ten minutes.

For the working of all the electrical household appliances that have been mentioned some source of electrical supply is, of course, required, and the best to adopt must depend upon the position of the house, its size, and the precise amount of duty that electricity will perform in it. The different bell-annunciator and alarm systems generally require surprisingly little power to operate them, and no difficulty will be found in supplying each system from a battery; or it may even be possible to let one battery suffice for all. Three or four cells of the Leclanché type will sometimes work a bell system continuously for two years without any attention, but it is always well to replenish a battery in time before its activity is exhausted, lest at some important moment it fail in its duty.

When, however, it is desired to supply a house with electric light, heat, or motive power, batteries for these purposes, unless on a very small scale, are hardly to be recommended. In the first place they would necessarily be troublesome to maintain, and in the second they would, in the present state of the art, be very costly. Power is obtained from a battery by the slow combustion of the zinc in its plates, and as metallic zinc is not found like coal, ready prepared in nature, the process for obtaining it is by comparison expensive. Probably no battery in existence can furnish electrical power at theoretically less than twenty-five cents per horse-power per hour in material alone, while actually the best cost, as a rule, thirty-five cents; and a horse-power is well applied locally if it gives illumination equivalent to two hundred candles. The only prospect that seems open to the extended successful application of the primary battery to light and power is in the possibility of its chemically producing, during its work, compounds which have directly or indirectly a commercial value. If this end, which has long been striven for, could be successfully attained, and a sufficiently large market found for the produce, the battery might come forward as the most advantageous source of electrical supply.

At present, however, recourse is had to mechanical sources of electricity —dynamo-machines driven by steam-engines—and it is no exaggeration to say that the practical success of electric lighting is due to the dynamo-machine as a source of electric supply.

In towns electric lighting from central stations is developing so rapidly that it is now very generally possible to obtain electricity from street mains

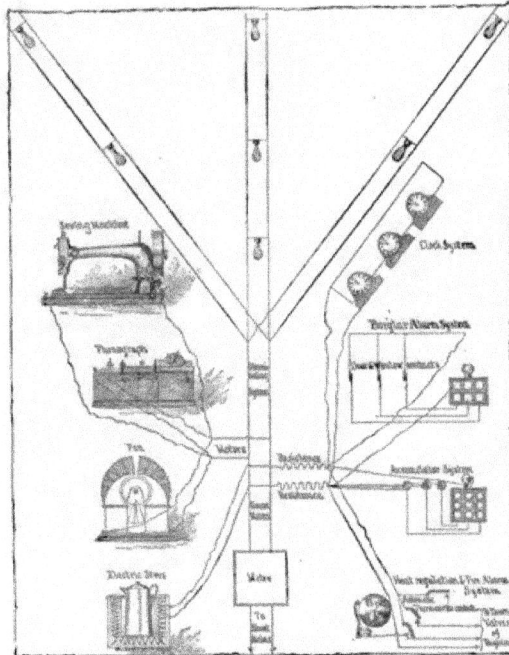

Plan of Wiring a House for its Various Electrical Appliances.

It will be seen that the house mains are connected through an electricity meter direct with the street mains ; that the lamps, heaters, and motors, wherever they may be situated, are operated directly from the house mains or their ultimate ramifications ; while the annunciator, clock, and alarm systems are all operated from sub-mains connected to the central mains through simple regulating coils of wire, termed resistances. Their object is to reduce the electrical pressure on these sub-systems to the desired limit for their effective operation, since the whole electrical supply they need probably does not exceed that given to one incandescent lamp burning continuously. In this way electrical economy is obtained and the safety of the more delicate apparatus insured.

like gas or water—a plan that will always be more economical and convenient than local production. Large buildings, or groups of

17

buildings, may sometimes be lighted advantageously by a local plant, but for a town dwelling a separate engine and dynamo is generally out of the question. The great convenience attending electrical supply from the street mains is the absence of all batteries and their attendant requirements. The arrangement of the different interior systems on this plan is illustrated in the accompanying diagram.

The diagram [p. 259] shows that the house mains receive their supply through a meter which keeps a register of all the electricity traversing it. The electricity on entering the meter has two paths open to it, one by wavy metal strips above, and the other through coils of wire and bottles beneath. The proportions of these are so arranged that the strips conduct, let us say, one thousand times better than the coils and bottles, and, as electricity always divides between two paths in the exact ratio of their conductivity, the quantity which passes through the strips will be just one thousand times as great as that passing through the bottles. These latter are filled with a weak solution of zinc sulphate, and each contains two zinc plates about one inch square. In obedience to laws discovered by Michael Faraday fifty-five years ago, the electricity which passes through this bottle from one zinc plate across to the other through the solution, causes a certain quantity of zinc to be dissolved from the plate it leaves, and the same quantity to be deposited on the plate it reaches—the quantity of metal so transferred bearing an invariable known ratio to the electricity that has passed. These bottles are removed and replaced every month, and the change of weight in the dried plates compared after the lapse of that term. The amount of zinc found to have been transferred then determines the quantity of electricity that has passed through the bottle, and one thousand times this quantity has entered the house through the metal strips during that time. There is thus no machinery to get out of order, no moving parts to clog, or friction to overcome, and with the bottles exchanged every month the meter itself is almost imperishable.

For country-houses beyond the limits of central-station distribution, electricity must be locally produced to furnish light and power. For installations not exceeding fifty lamps, the gas-engine advantageously replaces steam when coal gas is obtainable at a moderate rate, as there is then no boiler, and less attention is also required. A gas-engine and dynamo in a barn or neighboring outhouse form a very convenient electrical supply, and, as a matter of fact, a given quantity of gas so burned in a good engine not only expends its vitiating products of combustion out-of-doors, but will also yield from twenty-five to thirty per cent. more illumination through the medium of electricity than when furnishing light directly. That is to say, the electric lamp is so much more economical in energy that it gives this excess notwithstanding the necessary loss in the engine and dynamo inherent to the conversion of heat into electrical power.

The mistake is sometimes committed of scaling up wires in the plastering of walls, as though they were not liable to a mortality from which even the electricity they convey does not render them exempt. It is certainly best to have wires placed out of sight, but where access to them can always be had if needed; and generally, if an electric system is worth introducing into a household, it is worth carrying out, not lavishly, but thoroughly and well. It is also unfair to suppose that an appliance needs no supervision or repair because it is electrical. Every such instrument is essentially of mechanical nature and inevitably subject to the requirements our knowledge of mechanism leads us to expect. On the other hand, when proper care and judgment have been exercised upon the introduction of these appliances, the superiority of electricity for domestic purposes over every other known power (even in the matter of independence from supervision) is incontestably exemplified.

Nor is it to be supposed that any of the applications above alluded to are visionary, for all are in actual use. Some are still regarded in the light of luxuries, it is true, but almost all necessaries were once in that favored class. Even tobacco is regarded to-day as a necessity of existence, and if history tells truly, table knives and forks were luxuries of the most extravagant type two hundred years ago.

Considering, then, that the household is in itself the condensed history of a nation's past, the centre os its present, and the cradle of its future, it is doubtful whether, among the many triumphs of the age that electricity may claim, any can be quoted of brighter renown than the rapid progress it has already made in the cultivation of the arts of life, and its adaptation to the needs and graces of the household.

ELECTRICITY IN RELATION TO THE HUMAN BODY.

By M. ALLEN STARR, M.D.

EXTRAVAGANT CLAIMS FOR ELECTRICITY AS A CURATIVE AGENT—THREE FORMS OF ELEC-
TRICAL ENERGY—FRICTIONAL ELECTRICITY—ITS EFFECT ON THE HUMAN BODY—NO
CURATIVE POWER IN AN ELECTRICAL BREEZE—A GENTLE STIMULANT TO CIRCULATION
—ELECTRICITY AS A "MIND CURE"—VOLTAIC ELECTRICITY—CATALYTIC EFFECTS OF
A CURRENT ON THE HUMAN BODY—AN AID TO NUTRITION—NECESSARY PRECAUTIONS—
CATAPHORIC ACTION—HOW COCAINE MAY BE APPLIED—ELECTROTONIC EFFECTS—USE
OF AN INTERRUPTED CURRENT IN PARALYSIS—SENSATIONS PRODUCED—IMPORTANT
SERVICE IN LOCALIZING BRAIN FUNCTIONS—DEATH BY ELECTRICITY—A LESS OFFEN-
SIVE MODE OF EXECUTION THAN HANGING—FARADISM—WORTHLESS MAGNETIC AP-
PLIANCES—ELECTRICAL INSTRUMENTS—THE PROBE AND CAUTERY.

THE effect of electricity upon the human body has in recent years
become a subject of interest both to the general public and to
scientific observers. It is one which is continually forced upon the
attention by the reports in the daily press of accidents due to strokes of
lightning, or to contact with electric wires on the streets; by the recent adoption of electricity as a means for executing criminals : and by
the extravagant claims of the curative powers of electricity in disease.
The actual changes produced in the body by this form of energy, its real
effect in the treatment of maladies, together with the aid which electric
apparatus can render to physicians, have received careful investigation
both by physicists and physiologists. Even the insane are under the
mysterious spell, for witchcraft has given place to electricity in the deranged imagination, and it is the voice of the telephone which is now
heard by the lunatic who formerly complained of the suggestions of the
devil.

The general interest in the subject, therefore, has made it proper to include in the present volume one chapter upon the relation of electricity to the human body. And this may be of service not only by summing up what is definitely known as the result of recent scientific investigations, but also by clearing away some of the mysterious and erroneous assertions of those whose interest it has been to deceive the unwary.

In the preceding chapters it has been shown that electricity is one of several forms under which energy becomes appreciable, and that, like heat, light, or work, it is measurable, and can be produced by or converted into other forms of energy.

There are several manifestations of electrical energy which, though all one in their nature, must be distinguished from one another in their application to the body.

These are known as frictional or static electricity ; current or voltaic electricity, commonly termed Galvanism ; and induced electricity, or Faradism. Each of these forms of electrical energy is produced in a different way and has its peculiar effect upon the human organism. Each, therefore, must be considered by itself.

I. *Frictional* or *static electricity* is the form which is produced by the friction of bodies which are by nature in a different electrical state. If one walks across the room without lifting the feet from the floor, on a dry day, and then touches any metallic object, a spark flies from the finger to the object. Every one has experimented in lighting the gas in this manner. The effect is due to the fact that frictional electricity has been generated in the body by the rubbing of the feet on the carpet, and that the dry state of the air has prevented its immediate diffusion from the body into the surrounding atmosphere. If the air is damp such diffusion occurs so constantly that no accumulation in the body is possible, and hence on a damp day one cannot get a spark from the finger. The amount of energy in the spark is proportionate to the expenditure of energy in friction, for one will get a larger spark by walking twice across the room than by walking but once, and by rubbing the feet along the carpet instead of stepping lightly. The same effect can be produced by holding one's hands against a revolving glass ball, a method of obtaining frictional electricity pointed out by Hawksbee. Without friction no spark can be obtained, and unless the metal be touched as soon as the friction is made no spark will fly. This shows that ordinarily the body is not in a state of electrification, and that even after being rubbed it soon returns to its natural, indifferent state. The body cannot be kept in a condition of electrical tension because of the constant diffusion of electricity from it into the air or into the ground. Therefore a permanent storage of electricity in the body is impossible, unless it be carefully insulated. This can only be done by placing a person on a high chair with glass legs, and by having in the room some dishes containing sulphuric acid, which absorbs moisture from the air. Under these conditions, if the body be rubbed vigorously, frictional electricity will be produced and stored up, as it cannot escape. But the moment the person rises from the chair, or touches any object in contact with the earth, a spark passes and a slight shock is felt; the tension is relieved and the natural condition is restored immediately. Instead of generating the static electricity on the person by direct friction, it is more convenient to convey it to the body from a frictional machine under the same conditions. This is the common method of applying static electricity, and almost every one has at some time " taken sparks " from such a machine. While the sparks are passing, various sensations are perceived and effects noticed, but as soon as they have ceased there appear to be no permanent results from the application.

Almost all substances are capable of being put into a state of electrical tension by friction, but when charged they do not always act alike. The experiments described in the first chapter have demonstrated that there are two opposite electrical states produced in objects, one termed for convenience positive, the other negative ; and it was also shown that an object in one of these states attracts objects in an opposite state and repels those in its own state.

Thus, if a pith-ball hanging by a thread and attracted to the glass rod becomes, by contact, charged with positive electricity from the rod, it at once jumps away from the rod, being repelled from it because it is now in the same electrical state as the rod.   In the same way the human body, like the pith-ball, can be charged with either positive or negative electricity, and while insulated after being thus charged, it will attract things in an opposite state and repel things in a like state.

It is known that when the body is put into an electrical state by friction the hair rises and stands on end.   This is because the body and the hair, being in the same electrical state, repel each other, and as the hair is easily moved, it rises.   The same repulsion is felt by the air about the body, as it becomes charged by diffusion from the skin ; if the body is rapidly charged the movement of the air about it may be rapid and be felt as a breeze.   This is the same as the electrical breeze which flows from the points of an electrical machine, and which may be made to blow out a candle or turn a little vane.   There is no more curative power in the electrical breeze than there is in any draught of air, although the most absurd statements of its effects have been made.

But if the amount of the electric charge is too great to pass off thus easily into the air, an occasional spark will leap from the body to any near object which is not insulated ; and, in fact, if the experiment be performed in the dark, one may see innumerable little sparks coming from the hair, so that there appears to be an electric halo about the head.

Careful experiments have shown that frictional electricity resides on the surface of the bodies which are charged with it.   Thus, if an insulated metal ball be covered with two spherical disks and then charged with electricity, it is found that when the disks are removed they contain all the electricity, and the ball beneath them has none at all.   The same is true of the human body, and therefore frictional electricity never penetrates beneath the skin, or produces directly any effects upon the deeper tissues.   But it may produce indirect effects.   The skin is, of course, very sensitive; and any sudden change of electrical state produced in it, such as the giving of sparks, causes a decided irritation of the surface.   This is appreciated by pain at the point from which the spark jumps, and if the spark be large it may burn and raise a little blister. The degree of irritation depends upon the amount of electricity discharged from the point of skin. This may be so great as to burn seriously or even to destroy life, as is seen when one is struck by lightning. Now, any irritation upon the sensitive nerves of the skin, whether by a spark or a sudden blow, sets up a nervous impulse which is carried along the nerves and which causes a number of effects. One of these effects is a sudden movement of the irritated part. One suddenly draws one's hand away from a lighted cigar before one realizes what is touched. Another effect is the reddening of the skin, which is a sort of provision of nature to counteract the effects of any destructive process by increasing the nutrition of the part injured. And the third effect is a conscious perception of a sensation which leads to a train of thought and a state of emotion pleasurable and invigorating, or painful and depressing, as the case may be. The electric breeze is rather pleasant, while sparks are decidedly disagreeable. Hence, even though the direct effect of frictional electricity may be limited to the surface, its indirect effects may be general.

But the same kind of indirect effects may be produced by any mild irritation of the skin. The general effect of static electricity is, therefore, about the same as that of a cold bath, or the muscle-beating of the Swedes, the lomi-lomi of the Sandwich Islanders, the whipping with twigs in a Russian bath, the needle-douche of a Turkish bath, or any other sharp mechanical irritant. All these agents stimulate the circulation, and produce a sense of refreshment which is harmless, and may even be beneficial. There is no mystery about their action, and no very marked curative effect in disease. And there is no special curative power in static electricity which is not common to them all.

There is certainly something rather startling in having sparks applied to or drawn from the body. And this has led to the employment of static electricity to produce marked mental impressions. In the present day, when "mind cure," "Christian science," and "hypnotic suggestion," are discussed on all sides, it needs no argument to prove the interaction of mind and body. Any one whose toothache has left him at a dentist's door, or whose digestion has been deranged by anxiety, can testify to the fact. And the decided effect of expectant attention in modifying slight functional disturbances is admitted by every one, and may be honestly employed in the treatment of disease. It is not surprising that in some maladies an agent so startling and impressive as electrical sparks should be employed to excite the expectation of cure. In hysterical persons, whose ailments are due to a deranged imagination, it often suffices to impress the mind with a positive persuasion that the agent employed is able to cure, and the effect is obtained. How strong the effect of this persuasion may be is witnessed by the almost universal belief during the middle ages of the efficacy of the royal touch in the cure of the "king's evil;" to say nothing of modern miracles. If mystery can be invoked as an aid in the treatment of these imaginary affections the cure is more certain to follow, for the state of expectancy is heightened. It is not surprising, therefore, that the supposed mysterious powers of electrical sparks should have been extolled for the purpose of impressing the mind. Nor is it to be wondered at that without any deception favorable results should have followed the use of static electricity in the hands of physicians who can distinguish the class of cases in which it is likely to be successful. But such results are wholly indirect and due to mental expectation, and should not be ignorantly ascribed to the action of the electricity. And a knowledge of this fact should prevent the abuse of the agent, or any expectation of a curative power in real organic and serious disease—where it cannot be of service.

If the electrical condition of the body remains for a time different from that of the surrounding atmosphere, as it may on a dry day, it is supposed that the state of tension produces an indefinite feeling of discomfort. Such a sensation is often complained of by certain persons just prior to or during a thunder-shower. Some, indeed, are quite prostrated by the occurrence of an electric storm; and the susceptibility of nervous persons to changes in the weather has been ascribed to this cause, though it is often due to a peculiar reaction of the body to dampness rather than to electricity. That persons in a state of illness are more liable to notice such slight electrical changes in the atmosphere than those in health is a well-known fact, and it is certain that those who suffer from neuralgia are especially sensitive.

II. *Current* or *voltaic electricity* is different from static electricity in its mode of production and in its effects. It is produced by chemical action in a battery, the work done by the expenditure of energy in the chemical process being partly manifest by the electrical state produced in the elements. There is more or less chemical action going on constantly in the process of nutrition within the body, and therefore the body may be looked upon as a sort of battery for the production of electricity. But the amount thus produced is far too small to be appreciable except by the most delicate tests, and may be disregarded. Whenever, by chemical action, two elements are simultaneously put in a state of electrification it is found that their electric condition is unlike, and if they be joined together there is a tendency for the difference between them to be equalized by means of the passage along the line of connection of a so-called current. Thus, if a copper cent be laid on the tongue and a silver quarter placed under the tongue, their edges in contact, a current passes through the tongue, and is perceived as an acid taste. Or if pieces of zinc and of carbon be placed in a tumblerful of dilute sulphuric acid, and their free ends be joined by a wire, it is found that as the acid attacks the zinc a current begins to pass along the wire from the carbon to the zinc. And this current may be so intense, if the acid is strong or the zinc plate is large, as to attract the needle of a compass, or to heat the wire red-hot. It is not necessary to suppose that anything is actually running through the wire—as the term current might imply—but only that a change of state is taking place in the wire which tends to propagate itself in a definite direction from the carbon toward the zinc. Suppose that the wire be cut in two, its ends attached to sponges, and the sponges laid upon the body; then this change of state which takes place in the wire attached to the carbon up to its point of contact with the skin is propagated to the other wire through the body, and the parts of the body between the two sponges are put in the same state as the wire. It is then said that a current is passing through the body. And just as the wire was heated by the current, so the body may also be heated or otherwise affected if the current is sufficiently strong. Wire is found to assume this change of state easily, and hence it is said to be a good conductor. The body, however, does not, and hence it is said to resist the action of electricity, or to be a bad conductor. How bad a conductor it is may be judged from the fact that under the most favorable circumstances it offers as much resistance to the passage of a current as does three hundred miles of ordinary telegraph wire.* It is evident, therefore, that strong currents have to be employed in order to affect the body at all. It is found, however, that the chief resistance to the passage of electricity is offered by the skin,† which is practically a non-conductor, and in order to overcome this it is necessary to keep it warm and moist with a solution of salt, or else to penetrate it with needles and apply the electricity through them. The former of these methods is the one commonly employed, and even then the resistance of the body amounts to about two thousand five hundred ohms. The tissues beneath the skin offer resistance in different degrees, the muscles conducting much more readily than the nerves or the bones. It has been supposed that the electricity in passing through the body is generally diffused through the tissues between the poles. But there is no reason to believe that the human body acts like a homogeneous mass, and it is probable that electric currents uniformly pass along the lines of least resistance. There is every reason to believe, therefore, that when a current is sent through the body it is not uniformly diffused in the tissues, but passes chiefly through the muscles and blood-vessels, which offer the least resistance, and affects to a slighter extent the nerves, and least of all

---

* The resistance of the body is about two thousand five hundred ohms. A mile of No. 6 iron wire has a resistance of 8.54 ohms.

† The dry skin has a resistance of about one hundred thousand ohms.

the central nervous **system** (the brain and spinal cord), which **is** protected by its bony covering. In order to reach the nerves the current must be specially applied to them. An illustration will make this clear. If from a reservoir several **pipes lead** out, **some of** which are full of rubbish and others free and open, it is evident that much more water will flow through the open pipes than through **those which are** obstructed, and if it is necessary to wash the latter clean, some pumping apparatus is to be used, or by closing the other pipes **a great pressure** of water in the reservoir **must be secured.** The same thing **is true of** the passage **of** electricity through the body. It will pass along the easy ways, and as **it is** impossible to **concentrate** accurately its action upon any one set of tissues, its effects must **always** be uncertain.

The amount of electricity sent through the body is measured **by a** galvanometer* exactly **as currents in wire are measured.** But the body is so sensitive that only a few thousandths **of** an ampère of current can be safely borne. When it is remembered that several hundred **ampères** pass along the wires of an electric light **the danger of receiving a current** from them becomes at once evident.

When a voltaic current is passing through **the human body three** different effects are produced, which are termed respectively catalytic, cataphoric, and electrotonic.

(1) Catalytic effects. **A current of** voltaic electricity, when sent through a compound substance, decomposes it into **its elements, and** the action is termed electrolysis or catalysis. Thus, a current sent through water splits the water up into hydrogen and oxygen gases, the former of which may be seen coming off from the negative pole in bubbles. Now the human **body is a** highly complex structure, and, being affected by the **electric** current exactly as **water** is affected, it is decomposed by the passage **of** electricity. With weak currents this process may be but slight, and, since many nutritive processes are attended by such chemical changes, it has been proven **that a** mild electric current may aid nutrition by hastening or assisting the chemi-

*See "Electricity in the **Service of Man.**"

cal changes which are ordinarily going on. This effect of electricity in aiding nutrition has been cleverly shown by Professor Thacher, of Yale College, who applied the current for a week at a time alternately to the two arms of a person suffering from paralysis, and, by measuring the strength of the hands at the end of each week by an instrument which records degrees of power, he found that the power increased more rapidly in the arm to which the electricity had been applied than in the other arm. Thus the gain in size and strength under the use of Galvanism could be shown as follows:

	Galvanized arm.	Untreated arm.
(1) Gain in strength first week (left)..........	17°	12°
(2) Gain in strength second week (right)......	15°	10°
(3) Gain in strength third week (left)........	7.4°	0.9°
Total............................	39.4°	22.9°

That is, the galvanized arm made almost double the progress of the other (1:1.72). In the same person there was no evidence of gain from the use of Faradic electricity or from massage. Such results in aiding nutrition in the body are due to the increased rapidity of chemical changes caused by the catalytic action of electricity.

When, however, the current used is strong, the catalytic action becomes so intense that tissues are destroyed and blisters are raised at the two poles, the one at the negative pole being filled with alkaline, and the one at the positive pole with acid fluid. At the same time the tissues about these blisters become affected much as if they had been burned, and the resulting sores are slow to heal and leave deep, permanent scars. It is evident, then, that a strong electric current is very destructive to the body tissues.

This catalytic action is more intense the more concentrated the current, and therefore, in its medical application, small-pointed instruments are used. The action is employed for the decomposition and destruction of new growths, such as tumors, in the body, and in order to enlarge passages or cavities which have become constricted by disease. It is not, however, generally approved, because its action is not accurately limited and is somewhat uncertain; its after-effects are unpleasant; it is extremely painful; and other methods of surgical procedure are far more precise and successful. This catalytic action has also been employed to remove hairs from the face, a fine needle being inserted at the side of the hair and the current being allowed to destroy the root from which the hair grows. Here it has proved of service, and if carefully employed it leaves no scar.

Since a certain amount of catalytic action is always attendant upon the passage of a current through the body, the use of electricity must always be considered as injurious unless proper precautions are taken to avoid the strong currents which do harm. The reckless handling of electric batteries, or the giving of shocks by those who have no purpose excepting to amuse, is therefore to be avoided, and the danger of touching a wire used in electric lighting cannot be too strongly urged, as the strong current may disintegrate the body tissues so rapidly as to destroy life in a few seconds.

(2) The second action of a continuous galvanic current is to move along with it the fluids which lie in its path. This is called its cataphoric action.

If a partition of parchment be fitted in a bowl, and two fluids, one salty, be poured into the compartments, and if the positive pole of a battery be put in the salty fluid and the negative pole in the other, the electric current will carry the salty fluid through the parchment into the other compartment, and there the fluid will rise to a higher level. The passage of fluid through the parchment is called osmosis. Such an interchange of fluids through the membranes of the body forms an important part of the process of nutrition and of growth. As electricity will promote osmosis it may increase the nutrition of the parts to which it is applied.

A practical application of the cataphoric action of electricity has recently been made. It is well known that a solution of cocaine applied to many of the tissues makes them insensitive. If a drop be put in the eye the eyeball can be touched without causing a wink. But cocaine does not affect the surface of the skin, and to render this insensitive the drug must be applied to its under-surface. This is usually done by hypodermic injection. But it has been discovered that an electric current may be used to carry the cocaine through the skin, and thus render it insensitive. This is done by applying the cocaine on the positive pole or sponge of the battery and placing this over the part to be rendered anæsthetic. The success of this experiment proves that the galvanic current will carry fluid substances through the skin and into the body. This method, though ingenious, has not been widely employed, because it is more painful and less convenient than the method of hypodermic injection. It is used, however, to quiet the pain of neuralgia. Medicines have been administered in this way in medicated electric baths, but it has been found impossible to regulate the amount given, and hence they have fallen into disuse.

(3) The last effect of the voltaic current when passing is to produce under each pole a peculiar molecular state of the tissues termed electrotonus. Inasmuch as the condition at each pole is the opposite of that at the other, there are two kinds of electrotonus, named respectively anelectrotonus (anode, positive pole) and katelectrotonus (kathode, negative pole). The condition at the anode is one of diminished excitability; that at the kathode is one of increased excitability.

. This is the only effect of the electric current which continues after the current ceases to pass, and it may last for a considerable time.

It is a property of living tissue to be excitable to several kinds of stimuli, mechanical, physical, or chemical; and the excitability varies from time to time. There are times, for example, when one feels irritable and uneasy; there are other times, after a good meal, when there is a sense of comfort and repose. What is true of the organism as a whole is true of its constituent parts. And it is one of the powers of electricity to produce in the molecules which make up the various organs changes of irritability in one direction or the other. These changes are greatest in intensity near the poles and diminish at a distance from them, so that at a point in the body half way between the poles one effect is neutralized by the other.

It is well proven that this electrotonic effect is quite intense in its results upon the circulation, that it stimulates it, and thereby has a beneficial effect in promoting nutrition. To it are ascribed what are known as the refreshing effects of an application of electricity, the sense of invigoration and comfort that follows the use of a general mild current applied to the entire body when fatigued. Such an effect is quite comparable to the restoring effects of a swim on a hot summer day—and the latter is a less expensive luxury than the former.

To the quieting effect produced by the positive pole of the current is due the relief of pain which follows its application. For it is generally admitted that in many painful affections of a nervous or muscular character the application of a mild continuous voltaic current, with the pole upon the painful part, affords prompt relief.

The quieting or exciting effects of electricity would be of much greater advantage in the practical use of the agent, however, if they could be more exactly controlled. It is found, unfortunately, that after a current has passed through the tissues for a time and is then stopped that a condition ensues under the respective poles which is just the reverse of the former state. Thus under the anode, where excitability had been quieted, it is now greatly increased above the normal, a complete reaction having taken place. Hence the anode may fail to relieve pain permanently, though quite effective for a time. And under the kathode, where excitability had been increased, there is at first a quieting effect of very short duration followed by a renewed condition of irritability more intense even than the former state. Hence upon the cessation of passage of a current there remains at both poles a state of excitability. The passage of the current is therefore stimulant in its general and its local effects. It is this stimulant effect which causes the increased circulation already alluded to and the sense of invigoration. It is perceptible in the redness and heat of the skin under both poles, which remains for some time. And it undoubtedly has a decided and permanent effect upon the

18

elasticity of the walls of the blood-vessels, for if, several hours after an application of electricity, and long after any redness of the skin has passed away, the limb be placed in a warm bath, the redness at once returns in the parts to which the poles have been applied.

It is upon this electrotonic effect upon the tissues that great stress has been laid by those who employ electricity in the treatment of disease. It must be admitted that the effect is produced and that it has some duration. Whether it is of a kind to affect beneficially the various changes in living tissues which occur in the different forms of disease in which it has been employed is an open question, and one upon which the experience of physicians leads them to differ.

The interrupted voltaic current. It has just been stated that on stopping the current as it passes an increased state of excitement can be induced in the tissues. If it is desired, then, to produce excitement, the result can be reached by making and breaking the circuit alternately, thus producing an " interrupted current." Or, if a great increase in the irritation is sought, the current can be reversed rapidly, so that a part of the body may be thrown alternately into a state of anelectrotonus and katelectrotonus, each being used to reënforce the other. Such reversal of the current is termed a " voltaic alternative," and is most intense in its effects, producing more severe shocks to the body than any other procedure.

The excitement produced by an interrupted or an alternative current is employed very extensively in the treatment of paralysis, for by it muscles may be made to contract when they are incapable of voluntary exercise. By the same means impulses may be sent along nerves. It is the sudden change of state in the muscle or nerve caused by the electric shock which throws it into action. The passage of a continuous current does not have the same effect. The interrupted current, therefore, differs from the continuous current in its action on the various organs. It appears to have the power of setting them to work, and this effect is of considerable interest.

When an effort is made to close the hand there are several processes gone through with. There is an act of the will directing the character of the movement to be made. This takes place in the brain. It is transmitted along the nerves, from the brain to the muscles, as an impulse. Then there is a contraction of the muscles producing the desired motion. Now, each of these processes can be set going by an electric shock. Thus if the current be applied to that part of the surface of the brain which controls movements of the hand, the motion is made just as if by act of will. Or if the current be suddenly applied to the nerve in the arm, an impulse is there started which also travels to the muscle and causes the hand to close; or if the current be suddenly applied to the muscles of the hand directly, and not through the nerve, the hand again will close. It is evident, then, that the electricity will produce the same effect as the will-power, no matter whereabouts it is applied in the motor apparatus.

Sensations are also produced by the interrupted current. Tickling or numbness, or heat or pain, is felt at the point of application of the electric poles on the skin of the hand according to the strength of the current. If the current be applied to the nerve in the arm which comes from the skin of the hand, the same sensations are felt in the hand. This is because the hand is joined to the brain by nerve-threads, each part of the hand being joined to its own particular area of the brain, so that one can imagine a little map of the hand projected on the surface of the brain, and all sensations from the hand are perceived in this little area of sensory brain-surface. Now as all sensations reaching that brain-surface have in one's experience come from the hand, when a change of state is produced in that brain-area it is perceived as a sensation and referred to the hand. Hence when an impulse started by electricity in the nerve-threads from the hand arrives at the brain a perception follows, and the sensation is referred to the hand, the brain having no means of correcting its false impressions except through the aid of other senses. It is because of this fact that people have sensations in amputated limbs long after the amputation, and also for this reason that imaginary pain is really as severe as any other. Electricity, then, sent through a sensory nerve will produce changes of state in the brain which we know as sensations of touch, temperature, or pain.

The same is true of the sensory nerves of sight and hearing and smell and taste. If a current be sent through the eye we see a flash of light. If the experiment be conducted in a dark room, it will be found that when the negative pole is applied to the eye and the current is sent, a whitish-yellow light is seen; when the current is stopped, a bluish light is seen; when the positive pole is applied to the eye and the current is sent, a bluish light is seen; when the current is stopped, a whitish-yellow light is seen. Here, then, is an ocular demonstration, not only that electricity affects the optic nerve, but also that the poles differ, and that closing the circuit at the positive pole and opening the circuit at the negative pole are similar in their results. If a current be sent through the ear, a noise will be heard when the negative pole is put on the ear and the current is closed, also when the positive pole is put on the ear and the current is opened. The acoustic nerve lies so deep that these effects are only obtained by strong currents. When a current is going through the head at any point a decided metallic taste is perceived in the mouth, which is acid at the positive pole and salty at the negative pole. It is said that the olfactory nerve may also be excited by the current, which is then smelt as well as tasted.

All these experiments show that when a shock is given to a sensory nerve an impulse is sent to the brain and there produces a change of state which is perceived as a sensation. The kind of sensation received depends upon the part of the brain affected, since each part has its own function. The functions of the brain may therefore be set in action by electricity reaching it through the nerves.

But the same is true if the electricity be applied directly to the brain-surface. This direct application can be made in animals, and has been made in man when, by a fracture of the skull, the brain has been laid bare. Sensations of light, sound, touch, taste, and smell are then perceived, just as when they were excited by its application to the sensory organs or to the nerves from the organs to the brain.

The difference of effect when different parts of the brain are excited affords one of the many proofs that each part of the brain has its own particular work to do. And, as a matter of fact, the earliest experiments which led to the discovery of the localization of brain-functions were made by the application of electricity to the brain. Since this theory is now established, and is being practically applied in the detection and localization of brain-diseases and in the successful removal of brain-tumors, the discovery that interrupted electrical currents can set the brain in action has been one of real importance to medical science, and has resulted in the saving of human life.

Another important fact in this connection is one of more interest to physicians than to others. It is the fact that in certain diseases the muscles and nerves lose their power of being excited by rapidly interrupted electrical currents such as are produced in a Faradic apparatus. This fact is used by physicians as an aid to diagnosis. For it enables a paralysis due to disease in the nerves to be distinguished from paralysis due to disease in the brain. It also serves to expose those whose object it is to deceive. Lazy soldiers in the army, lazy criminals in prisons, and chronic loafers in hospitals, who do not want to work, often feign paralysis and plead inability to move, and by this test it is possible to expose the deception.

During the last few years, since electric lighting has been introduced generally, the newspapers have published from time to time accounts of serious accidents due to electric shocks received from the wires. In such cases the wire, which has not been properly insulated, has come in contact with some part of the body, and the electric current has passed through the body into the earth. The currents used in lighting are several hundred times greater than those which can be safely applied to the body. Therefore the shock received is enormous, and the sudden change of state in the molecules of the body is so severe as to arrest all vital processes. Very serious results are caused by such shocks. Usually the individual is killed instantly. Sometimes he is only stunned for a time, and then he becomes delirious, and subsequently he is found to have paralysis or blindness or epilepsy or even insanity. The most serious forms of nervous disease may be produced by such a sudden change of state in the molecules of the body. And the fact that it is a molecular effect is proven by the lack of any evidence of disease in any of the organs when examined with the microscope after death. The shock seems to suspend all vital activity in the finest cells without lacerating or destroying the different organs.

When the shock is received from a continuous current and this passes for a few moments through the body the tissues may be severely burned, and sometimes hemorrhages are found in the brain and other organs. When it is the alternating current which kills, no such effects are found. But as the shock in the latter case is more severe, death is more common in accidents from alternating currents.

The fact that death may be instantly caused by a severe electric shock has led to the proposition to use electricity in capital punishment, and it is well known that a commission of experts, after careful experiments, have indorsed this method, which has accordingly being adopted in several States. The method employed in the experiments in Mr. Edison's laboratory is to attach the two wires to any part of the animal's body, and then by closing the circuit, either with an ordinary key, which may be moved by the hand or by clock-work, or by a hammer, to send the current through the body of the animal. It was found that an alternating current of seven hundred and fifty volts was sufficient to kill a horse weighing 1,230 lbs. instantaneously, death being apparently painless. There can be no question that such a method of execution is more certain,* more sudden, and less offensive to spectators than that of hanging, and therefore it should be universally adopted in criminal execution.

III. The third form under which electricity appears is the *induced*

---

* That the first attempt to put the law into effect resulted in a horrible spectacle does not invalidate the statement made in the text. It merely demonstrates the necessity for a more careful construction of apparatus under the direction of electrical experts. Much more revolting scenes have occurred many times at executions by other methods.

or *magnetic current,* known as **Faradism.** The interruption in a very weak voltaic current is employed to produce a strong induced current in a coil of wire,* and this induced current is then applied to the body. It does not appear to have any catalytic or cataphoric action. Nor can it be said to have either an anelectrotonic or a katelectrotonic effect, because there is really an alternation of the two as in the voltaic alternative. It has about the same effect upon the body as a voltaic alternating current, and may be employed, as this is, to stimulate the nerves and muscles, to increase the circulation, and possibly to aid nutrition generally.

The induced current can be produced in an apparatus much smaller and more easily portable than any other current. Hence, whenever such effects as it will cause are desired, it is the one employed as a matter of convenience. Its limitations, however, are many, and hence it is less frequently used for medical purposes than the voltaic current. The fact that this current is induced by magnetism should not be taken to imply that by it magnetism can be made to act on the human body. The most careful experiments have shown that the human body is as completely insensitive to magnetism and as wholly unaffected by it as a piece of rubber or of wood. A person may stand between the poles of the strongest magnet, one which might hold up a ton of iron, without the slightest perceptible effect upon any of the bodily functions being produced. Hence all so-called magnetic appliances, brushes, or combs, disks, belts, and magnets have absolutely no curative power whatever. A few of these may, by friction, produce static electricity. Some are so constructed of two kinds of metal that on contact with the skin, whose perspiration is acid, a very weak voltaic current is produced, scarcely sufficient, after several hours, to redden the surface. The majority of the effects produced by such contrivances are due, like those of the static current, to expectant attention rather than to any action of the agents, which careful investigations have shown to be inert.

* For the means of production of the induced current the reader is referred to the first chapter.

The writer once demanded of the agent of a widely advertised "electric belt" some proof that an electric current was produced by it, and suggested that any such current could be detected by means of a galvanometer. This test was objected to, but a little frame holding a dozen pocket-compasses was at once produced, and the power of the belt, which doubtless contained some iron plates, to attract the needles of these compasses was shown as proof of the production of electricity in the belt. Probably many who were ignorant of the difference between electricity and magnetism had been deceived by this so-called test.

Not infrequently there is seen upon the streets an electric machine with ringing bells, and a large index adorned with a sign that "electricity is life." The man in charge invites the passer-by to test his power of taking electricity, and to any one who stops he relates marvellous stories of the strength displayed by some one who has just gone on, and the wonderful cures which he has made. Such machines are usually Faradic batteries, and when one takes hold of the handles the current is strong enough to cause the hands to close forcibly upon them, so that he cannot let go. The current can, of course, be increased up to a painful point. The strength of the current will be determined by the distance to which the outer coil of wire is moved over the inner coil, as in any Faradic battery. The index in such a machine has no relation to the strength of current, and is manipulated by a screw in the hands of the exhibitor. It is therefore a fraud, and does not indicate the strength of current used. There is really no limit to the amount of current which one can endure except a limit in the pain which can be borne. Currents received from such a machine or from any battery in this manner can never be specially beneficial, as they only produce a local effect upon the hands and arms, and when these are paralyzed the current should be applied to the weak muscles.

While it is evident, from the review of the various effects of the different forms of electricity upon the body, that some of the effects are powerful, it is also evident that they are only beneficial in so far as they increase nutrition. The curative powers of electricity are really very limited, and have often been exaggerated by those to whose interest it is to urge them.

Certain electrical instruments have recently been introduced into medical use which deserve mention. The electric light is so brilliant, and yet can be produced in such a small space without the danger of burning, that it has long been used for illuminating purposes by physicians. It can be introduced into the mouth, or even into the cavities of the body, such as the stomach, enabling the observer to see objects otherwise invisible. Its light can be reflected into the eye or ear or nose very conveniently, since the source of light can be moved.

Another electric instrument is the micro-telephone or phonoscope, by which very slight sounds may be detected and magnified. The sounds of breathing and of the heart may thus be made audible; also the sound produced by the contraction of a muscle, and even the sound of the pulse in the wrist may be heard. Such instruments are, however, too delicate and require too nice adjustment to be generally used. A sound, however, having been recorded by the phonograph, may be subsequently made audible to a large number of persons through the graphophone, and this method has been employed by teachers in the instruction of medical students. The phonograph has also been used to record the sounds of coughing. Coughs vary in sound in different diseases, and by producing all the different coughs from the graphophone students may be taught to recognize the differences.

An electric probe or explorer has been invented for the detection of bullets in the body. There are several kinds of these explorers. Some are introduced into a wound like a probe, and consist of two fine insulated wires bare at the tip and connected with a battery. When the tip comes in contact with the metal bullet the electric current is completed between the wires, and a bell is rung, or index moved, by the battery. Others are constructed on a different principle. The presence of a metallic body near a delicate induced current is enough to change that current slightly by magnetism. An instrument has been constructed which will indicate the presence of a bullet deep in the body when the instrument approaches the right place. Such an instrument was used in the case of President Garfield, but it is said to have misled because of the magnetic effect of the metallic bed-springs upon it, a source of error which had not been considered.

Electrical instruments are also used to detect very slight differences of temperature in the body—a change of one hundredth of a degree Fahrenheit being easily shown by a thermo-electroscope. As such minute changes of temperature are constantly going on in health and are of no importance whatever for diagnostic purposes in disease, this instrument has been discarded, though charlatans have employed it to impress the wondering public by claiming to detect "congestions" which were imperceptible to the thermometer.

Electrical engines or small electro-motors have been employed by surgeons to run small circular saws for cutting bone, and in very delicate operations, such as those about the bones of the face or nose, they have been found very useful. Dentists use these motors in this manner constantly. The little saw can be held like a pen in the fingers and revolved at a high rate of speed by the electric motor, to which it is joined by wires.

The electric cautery is also widely used by surgeons, a delicate wire being heated to a white heat and touched to the spot which is to be burned.

These are some of the practical adaptations made in the construction of medical instruments, and in the future many other useful devices may be expected.

INDEX.